CW00369684

GREAT VOYAGES OF THE WORLD

25 adventures on the high seas and great rivers of the world

AA Publishing

GREAT VOYAGES OF THE WORLD

Contributors: Gary Buchanan, Elizabeth Cruwys, Ben Davies, Fiona Dunlop, Peter and Helen Fairley, Bob Headland, Christopher Knowles, Shirley Linde, Beau Riffenburgh, Anthony Sattin, Ann F Stonehouse, Rob Stuart, Mary Tisdall, Angela Wigglesworth

Produced by AA Publishing

Published by AA Publishing (a trading name of Automobile Association Developments Limited, whose registered office is Norfolk House, Priestley Road, Basingstoke, Hampshire RG24 9NY; registered number 1878835)

ISBN 0 7495 1585 6 (hardback); 0 7495 1967 3 (softback)
A CIP catalogue record for this book is available from the British Library.

Copy editor: Rebecca Snelling

Designer: Tony Truscott Designs

Colour separation by Daylight Colour Art, Singapore

Printed and bound in Italy by G Canale & C Sp A – Turin

Contents

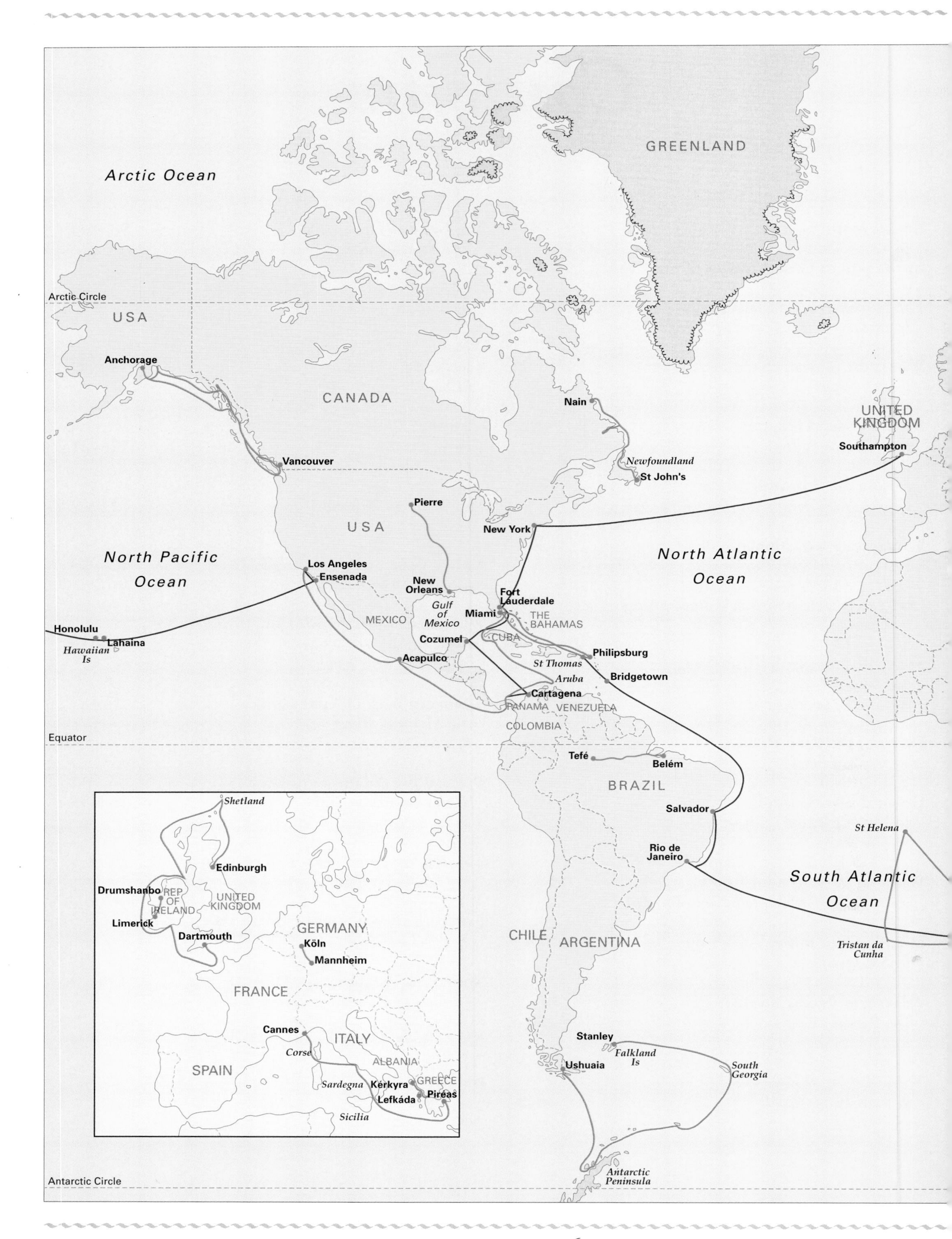

Arctic Ocean
GREENLAND
Arctic Circle
USA
Anchorage
CANADA
Nain
UNITED KINGDOM
Southampton
Vancouver
Newfoundland
St John's
Pierre
USA
New York
North Pacific Ocean
North Atlantic Ocean
Los Angeles
Ensenada
New Orleans
Fort Lauderdale
Gulf of Mexico
Miami
MEXICO
THE BAHAMAS
Honolulu
Lahaina
Hawaiian Is
Cozumel
CUBA
Philipsburg
Acapulco
St Thomas
Bridgetown
Aruba
Cartagena
PANAMA
VENEZUELA
COLOMBIA
Equator
Tefé
Belém
BRAZIL
Salvador
St Helena
Rio de Janeiro
South Atlantic Ocean
CHILE
ARGENTINA
Tristan da Cunha
Stanley
Falkland Is
Ushuaia
South Georgia
Antarctic Peninsula
Antarctic Circle
Shetland
Edinburgh
Drumshanbo
REP. OF IRELAND
UNITED KINGDOM
Limerick
Dartmouth
GERMANY
Köln
Mannheim
FRANCE
Cannes
ITALY
Corse
ALBANIA
SPAIN
Sardegna
Kérkyra
GREECE
Lefkáda
Piréas
Sicilia

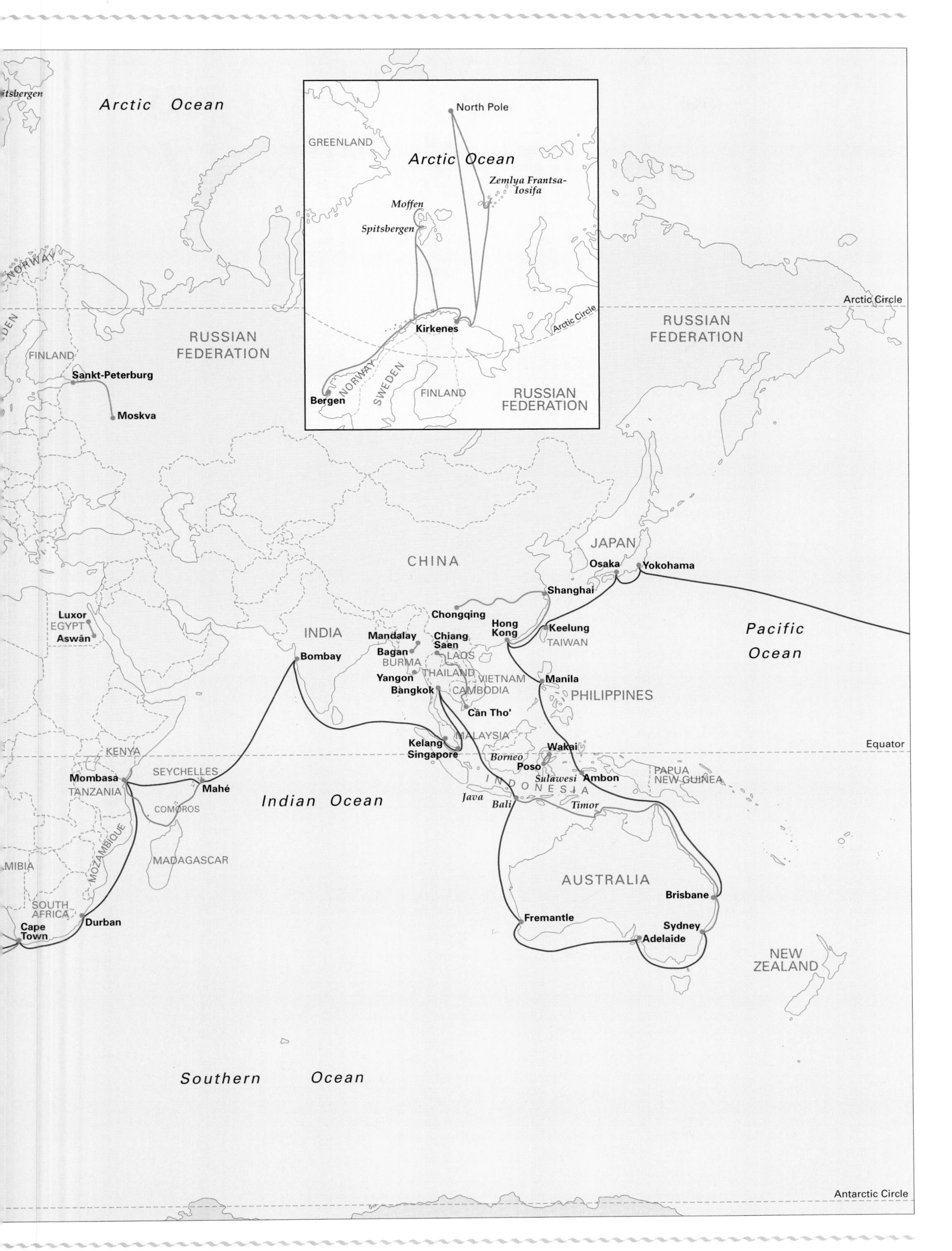

Arctic Ocean
North Pole
GREENLAND
Arctic Ocean
Zemlya Frantsa-Iosifa
Moffen
Spitsbergen
Kirkenes
Arctic Circle
NORWAY
SWEDEN
FINLAND
Bergen
RUSSIAN FEDERATION
RUSSIAN FEDERATION
FINLAND
Sankt-Peterburg
Moskva
RUSSIAN FEDERATION
Arctic Circle
JAPAN
Osaka
Yokohama
CHINA
Shanghai
Chongqing
Hong Kong
Keelung
TAIWAN
Pacific Ocean
Luxor
EGYPT
Aswân
INDIA
Bombay
Mandalay
Bagan
BURMA
Chiang Saen
LAOS
THAILAND
Yangon
Bangkok
VIETNAM
CAMBODIA
Cân Tho'
Manila
PHILIPPINES
MALAYSIA
Kelang
Singapore
Borneo
Wakai
Poso
Sulawesi
Ambon
INDONESIA
Java
Bali
Timor
PAPUA NEW GUINEA
Equator
KENYA
SEYCHELLES
Mombasa
TANZANIA
Mahé
COMOROS
Indian Ocean
MOZAMBIQUE
MADAGASCAR
AUSTRALIA
Brisbane
Fremantle
Sydney
Adelaide
NEW ZEALAND
SOUTH AFRICA
Cape Town
Durban
Southern Ocean
Antarctic Circle

ABOVE, *an early poster for the sleek cruise ships of the Cunard Line*

Great Voyages of the World

ON A SUMMER AFTERNOON IN 1843, a handsome, six-masted, black-funnelled ship eased out of her dry dock in Bristol and headed down the River Avon towards the sea. She was SS *Great Britain*, designed by the brilliant engineer Isambard Kingdom Brunel, whose intention was to revolutionise ocean travel. His remarkable ship had an iron rather than a wooden hull, was powered by a massive steam-driven propeller and had elegant, comfortable cabins that aimed to make the Atlantic crossing a pleasure rather than an ordeal.

But, for all her stateliness, SS *Great Britain* was little more than a yacht compared to the slendid liners that were to follow. Floating palaces such as *Mauretania*, *Queen Mary* and *Queen Elizabeth* allowed passengers to dine in sumptuous luxury, relax in smart nightclubs and cinemas, and even to take exercise on a mechanical horse.

Once travel on water did not necessarily equate with endless days of pitching and rolling in crowded, unsanitary conditions with poor food and stale water, people began to take journeys for enjoyment, rather than simply as a means of getting from one place to another. The era of passenger cruising had been born.

FROM SAIL TO STEAM

Although cruise ships did not appear until the 19th century, water-travel is almost as ancient as human civilisation itself. Peoples living on rivers built boats to ferry themselves and their goods from place to place – first of all on simple rafts made by tying logs or bundles of reeds together, then by dug-outs and later by canoes. Galleys appeared in Egypt about 5,000 years ago, while the Romans developed fleets of fighting ships in the Punic Wars (264–146 BC).

Some of the earliest travellers were pilgrims, who wandered the medieval world visiting sites of religious importance, while trade was always an important reason for undertaking journeys. Later, expanding empires developed mighty sailing ships to conquer and subdue new lands, and powerful navies evolved to protect their imperial interests.

BELOW, *SS* Great Britain *at her moorings in Bristol*

The development of steam power was a great advance in the history of ship-building. A few attempts had been made to run a boat by steam in the 1780s, but it was not until 1838 that the first steamer crossed the Atlantic – although allegedly its captain only made the journey by urging his mutinous crew to work at gun-point.

In 1845, after SS *Great Britain* delivered her passengers from Liverpool to New York in just 14 days and 21 hours, shipping magnates became convinced that steam was the way forward. Samuel Cunard, a Canadian from Nova Scotia, gradually converted his fleet of wooden sailing ships to splendid express 'liners' (ships that sailed a regular 'line' between two ports), including the magnificent *Lusitania* and *Mauretania*. His British rival, the White Star Line, built the luxurious *Titan* and her sister ship, the ill-fated *Titanic*.

Undaunted by the tragedy of *Titanic*'s maiden voyage, Cunard's company merged with White Star and went on to build the extravagant *Queen Mary* (1936) and *Queen Elizabeth* (1940). Cunard's *Queen Elizabeth 2* (1969) is still in service, and is generally considered to be one of the finest ships afloat, as well as having the distinction of being the last liner to be built. Seldom has any traveller been so cosseted in splendour as aboard *QE2*, with its glorious state rooms and promenades and daily entertainment that once included a Turkish bath, regular symphony concerts, a shopping arcade, gambling and a newspaper printed by the ship's own presses.

Ships at War and Peace

Six years before Cunard founded his business (in 1840), the Peninsular Company was formed to take passengers from London to Gibraltar. After the Suez Canal was opened in 1869, voyages were offered to India – the jewel in the British Empire's crown – and the company became the Peninsular and Oriental Line, now abbreviated to P&O.

Most of Cunard's and P&O's splendid ships saw action during the world wars, mainly as troop carriers, but World War II sounded the death-knell for the great age of liners as the technology which developed jet engines for fighter planes was applied to commercial aircraft. With planes crossing the Atlantic and flying to the most distant parts of the world in a matter of hours, only those people with plenty of time and money to spare could afford to spend days ploughing through the waves in a ship: the golden age of the liner was over. However, there was still a demand for the luxurious ships that cruised the turquoise seas of the Caribbean and the Mediterranean and the tradition of luxury and opulence was not forgotten; many modern cruise ships are virtually floating cities – one of which even has a short road lined with shops and a car that drives up and down it! In the 1970s and 1980s there developed a different kind of travel: for the first time, ships headed for the frozen seas of the Arctic and the Antarctic, showing passengers the majestic, dangerous beauty of lands that cost many an explorer his life.

Sold Down the River

The development of ships was not confined to use on oceans; rivers, too, offered travellers an exciting range of experiences. The wealthy of the late 19th century could embark on a three-week trip up the Nile, from Caïro to Aswan, stopping off at important archaeological sites such as Luxor and Karnak. One of the first of these trips was organised by the tour company Thomas Cook, which owned a number of the handsome hotels along the river. In the 1990s, Thomas Cook is still offering cruises on the Nile, although a holiday will cost considerably less in real terms now than it did in the 1890s.

Also steeped in history is a voyage up the gorgeous River Shannon in Ireland, following the route taken by St Brendan the Navigator in the 6th century, or a trip past the splendid castles and palaces overlooking the mighty Rhine as it meanders through Germany. For travellers more interested in the present than the past, a trip along the Mekong River through Laos, Cambodia and Vietnam offers a fascinating insight into the lives of the people who live in the beautiful countryside that was once ravaged by wars, while the endless rice-paddies along the Yangtze River provide a unique view of rural China.

Seas of Plenty

The variety of cruises available today is immense and many places are no longer inaccessible to all save the very wealthy. *Great Voyages of the World* aims not only to describe just a few of the experiences of people who have travelled on boats and ships all over the globe, but to provide enough data for readers to plan custom-made journeys for themselves. We hope your journeys will be as enjoyable as ours have been, and wish you good seafaring weather in the traditional manner of the US Coastguard: fair seas and a following wind.

ABOVE, *local guides can prove invaluable to the intrepid traveller*

LEFT, *the vast expanse of the mighty Amazon stretches as far as the eye can see*

The Norwegian Coastal Voyage

ELIZABETH CRUWYS AND BEAU RIFFENBURGH

ABOVE, *the traditional sight of clipp fish drying in the Arctic Circle*

From the dramatic, snow-capped fjords of Bergen, past the brightly coloured houses of tiny fishing villages to the frigid Arctic peaks of Spitsbergen, the Norwegian Coastal Voyage – known as the Hurtigruten – is not called 'the world's most beautiful voyage' for nothing. Round trips from Bergen to Kirkenes take 11 days, while the Spitsbergen excursion takes an additional nine days. The journey will take you to glaciers and cathedrals, waterfalls and museums, mountains and remote fishing settlements. Life on board the Hurtigruten ships is relaxed and luxurious, with excellent food and comfortable cabins looking out on to some of the most magnificent scenery in Europe.

BELOW, *onetime capital of Norway, the ancient city of Bergen offers dramatic scenery, history and culture*

Few cities can boast a setting as spectacular as that of Bergen, one of Norway's most ancient settlements. It is sometimes said that Bergen children are born wearing raincoats, and it was certainly pouring when we arrived, low clouds swathing the tops of the mountains with misty turbans. But the following morning was glorious: the water in the harbour a vivid royal blue, the hillsides above the city a rich, lush green.

Bergen has a great deal to offer the tourist, from 12th-century St Mary's Church and 16th-century Rosenkrantz Tower to the charming wooden houses along the *bryggen*, the main road along the harbour. A catastrophic fire in 1702 destroyed much of the town's medieval heritage and most of what escaped the first time round was caught in a second inferno in 1916. However, all has been faithfully restored and details of the city's history can be found in Bergen's wealth of museums: the Hanseatic and Shipping museums tell of the city's maritime past, while the Historical and Bryggen museums provide more general information.

ALL ABOARD FOR TROMSØ

The Hurtigruten was founded in the 1890s to provide an express boat service between Hammerfest and Trondheim. Fishermen in Hammerfest, who were used to letters taking five months to reach them during the winter, could now expect them in a few days and the first voyage in July 1893 was said to have been accompanied by cheers from all along the coast. From its very conception it was believed that tourism would be an important aspect of the Coastal Express and in the early years of the 20th century brochures were printed advertising the wild and rugged Norwegian coastline. Now, as the century draws to a close, the Hurtigruten fleet has 11 ships working between Bergen and Kirkenes, while the much smaller M/S *Nordstjernen* travels to Spitsbergen. All 12 ships have comfortable cabins, spacious dining rooms and plenty of observation lounges –

The journey starts in Bergen, Norway's second largest city and the gateway to the fjords. Founded in 1070, it nestles under towering mountains and its picturesque harbour and bryggen *(trading wharf) are well worth a visit. The first day actually begins at 10.30pm and, through that night, the ship passes hundreds of tiny islands and the lighthouse at Hellesoy, arriving at the art nouveau city of Ålesund at noon the following day, and at Molde, the 'City of Roses', later.*

Day three allows time for exploring the ancient city of Trondheim with its famous Nidaros Cathedral. The ship leaves at noon, and steams past some 6,000 islands and skerries, one of which is Monkholmen with its ancient abbey. During the night the ship crosses the Arctic Circle and enters the land of the midnight sun. Bodø and the rugged Børvasstindan Peaks are passed the following day, and the wild Lofoten Islands are reached in the early evening.
On the fifth day, time is allotted in Tromsø, a lively city with a polar museum and an Arctic cathedral. Next on the agenda is Hammerfest, the centre of the Arctic fishing industry and home of the Polar Bear Club. At Honningsvåg, it is possible to ascend the North Cape plateau.
Finally, on day seven, the ship reaches Kirkenes, a mining town near the Russian border.
Although the return journey makes some of the same stops, the timings mean that more sightseeing is possible – Berlevåg, with its spectacular views towards the Barents More, the gorgeous green-blue waters of Risøyrenna Channel, and Hitra Island with its huge herds of deer. On the afternoon of the twelfth day the ship reaches Bergen. For those taking the Spitsbergen option, the Hurtigruten ship is left in Tromsø and the Nordstjernen *is boarded, which passes the remote island of Bjørnøya on its way to the main city of Spitsbergen, Longyearbyen. Cruising along the rugged western coasts, visitors might see walrus hauled out on Moffen Island and will definitely see plenty of birdlife at the scientific station at Ny-Ålesund. After several days in the vicinity of Spitsbergen, the ship heads south toward Honningsvåg.*

ABOVE, *Ålesund, spread over three islands (two linked by bridge), is Norway's main fishing port*

although in July competition among some passengers for the best seats was intense.

We sailed out of Bergen at 10.30pm, with the last rays of golden summer sunlight slanting across the city and sparkling on the waters of the harbour. The route passes the island of Runde, Norway's famous bird sanctuary, where thousands of birds soar and swoop through the air.

The first major stop is Ålesund, sited on three separate islands and one of the prettiest fishing towns in Norway. Passengers are allocated about three hours on shore, enough time to wander through the winding streets and soak up the atmosphere. That evening, the ship docked at Molde, famous for its roses and its jazz. Early next morning the ship docked at Trondheim and we had until noon to explore this fascinating city that was once the capital. Founded in the 990s by Olaf I, it has a number of medieval buildings, including the Nidaros Cathedral (enlarged and rebuilt through the centuries) and the Archbishop's Palace (1160). The handsome Stiftsgården, a country mansion used by the Norwegian royal family, was built in 1770; it is said to be the largest wooden structure in Norway.

During the afternoon the ship eased in and out of some of the 6,000 tiny islands and skerries that litter this part of the coast, most of them swarming with birds. The Arctic Circle was crossed early the following morning and the ship then followed an impossibly complex route through narrow channels and past the towering Børvasstindan Peaks to Bodø, a town that was almost completely destroyed during air raids in 1940. Since rebuilt, Bodø now provides an important link to the remote Lofoten Islands to the north.

There are more than 80 islands in the Lofoten chain, many of them little more than inhospitable rocks that jut out of the sea like broken teeth. Here and there are emerald-green valleys but, for the most part, the islands comprise dramatic cliffs and craggy mountains. Little villages cling to their shores, some of them founded in Viking times by hardy fishermen who made a living from the rich cod stocks. While the human grip on these windswept islands might seem tenuous, birds are wholly at home. Puffins, kittiwakes, oystercatchers, terns, fulmars, guillemots, and even sea-eagles, swoop and soar in such numbers that they are impossible to count. Dividing the Lofotens from the bird sanctuaries of Værøy and Røst is the mysterious Moskenstraumen, a turbulent and unpredictable whirlpool. Fortunately, our ship keeps its distance.

Harstad is one of the largest of the fishing towns in the Lofotens and when we arrived its tiny resident population (some 21,000 people) was swollen by visitors who had come for the International Deep-Sea Fishing Festival that is held each July. Its quays were crowded with jostling fishermen, some displaying the largest fish we had ever seen outside an aquarium, and all proudly showing off their immaculate boats.

After Harstad, we set sail for Tromsø, the capital of the Arctic with two intriguing cathedrals and a fascinating Polar Museum, where some of us disembarked for an extra adventure – Spitsbergen.

NORTH TO THE ARCTIC

On first sight the *Nordstjernen* (built in 1956 and weighing in at 2,193 tons) seemed to us a little small to be braving the hostile Arctic seas, but a larger ship would not have been able to inch into some of the secluded fjords that characterise this part of the world and a great deal would have been missed. Spitsbergen means 'land of the pointed mountains' in Norwegian, an apt name for this wild and rugged island. It lies some 400 miles (640km) north of Norway, approximately 600 miles (960km) from the North Pole. High above the Arctic Circle, visitors between April and August will see the midnight sun.

We departed from Tromsø on a Saturday evening. Gulls dipped and swooped in our wake as we entered the open sea, the ship rocking gently at first and then with more vigour as the swell grew larger. By the time we reached Bjørnøya the next day only the hardy few were regularly standing at the rails outside, most passengers preferring to watch from the warmth of the lounge. Bjørnøya – Bear Island – was discovered by Willem Barents in 1596 and was named after a polar bear that was

ABOVE, *the wharf at Trondheim, one of the town's many attractive areas*

BELOW, *the jagged outlines of the Lofoten Islands, a region teeming with cod and birdlife*

killed by an expedition member. It is a desolate, fog-shrouded place with sheer cliffs and thundering surf, its beaches littered with the bones of walruses killed in the 18th and 19th centuries.

RIGHT, *11 huge concrete structures form the backbone of Tromso's ultra-modern Arctic Cathedral*

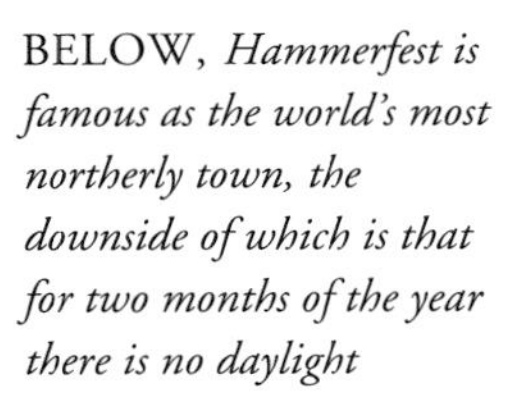

BELOW, *Hammerfest is famous as the world's most northerly town, the downside of which is that for two months of the year there is no daylight*

The next day we reached Longyearbyen, the capital of Spitsbergen and a frontier settlement that was founded in 1906 by an American named John Munro Longyear. Located on the Adventfjorden, it was essentially a mining village established to rip the rich deposits of coal from the surrounding mountains. The town is still dominated by mining and much of the land around it is owned by Store Norske Spitsbergen Kulkompani. Nearby Adventdalen is a wide glacial valley, the bottom of which turns into a vivid carpet of colour when hardy plants bloom in the short summer. We didn't have to walk far before the only sounds we could hear were those of the Arctic – the whistle of terns, the distant cackle of geese and the wind buffeting at our faces.

Back on the ship, we cruised along the vast Isfjorden, stopping at Barentsburg, a Russian mining settlement founded in 1920 and largely destroyed in 1943 by the German battleship *Tirpitz*. It was rebuilt in 1948 and, along with Pyramiden, marks Russia's presence in the

Svalbard archipelago. We then headed north to the 80th parallel and the nature reserve of Moffen, where lumbering walruses poked their leathery heads out of the water and watched us with their small, beady eyes.

The next day saw us at Ny-Ålesund, yet another old mining settlement. A series of accidents between 1929 and 1962 in which 70 people lost their lives led to the closure of the mine and the community is now a scientific site. Ny-Ålesund is a fascinating place. Once away from the wooden huts, where the scientists study the weather, ice and wildlife, there are huge empty plains fringed with snowy mountains. Flocks of geese gather each summer and large areas around Ny-Ålesund are off-limits to tourists so that these birds can breed in peace. One curiosity is the lovingly restored steam engine, said to be the most northerly train in the world, while near by there is a metal bust of Roald Amundsen, the Norwegian explorer who set off from Ny-Ålesund on his journey to the North Pole in an airship in 1926.

BACK ON THE HURTIGRUTEN

After another evening in Longyearbyen, *Nordstjernen* departed for the south, meeting the Hurtigruten again at Honningsvåg after an excursion up the dramatic North Cape, a vast slate plateau rising 1,008ft (307m) above the sea and forming one of Europe's most northerly points. Visitors can stand on the headland and watch the land drop off into the heaving sea and there is a precarious footpath leading to the actual most northerly point, which stops, rather alarmingly and very abruptly, at the edge of a sheer cliff.

After the fjords and mountains of southern Norway and Spitsbergen, the scenery of Finmark seemed rather tame. For the most part, it was swathed in an eerie mist so that only the merest glimpses of islands and hills could be seen. Kirkenes, the turning point of the voyage, is a centre for metal ore exports. Its position near the Russian border led to some serious fighting during World War II: visitors can stand near the border and look from the scrubby landscape of northern Norway across to the scrubby landscape of northern Russia!

The southbound journey was every bit as wonderful as the voyage north. Hammerfest claims to be the most northerly town in the world, having been given town status in 1789. Sadly, fires in 1890 and during World War II destroyed most of its historic buildings. It was the first town in Europe to be given electric street lighting (1891), an amenity not to be casually dismissed considering that daylight is completely absent between 21 November and 23 January. The residents' gardens are a great attraction to reindeer and one irate housewife told us that as their pastures become bare they flock to Hammerfest for fast food, stripping vegetable plots and causing traffic jams.

The Hurtigruten ship hits Tromsø at about midnight, which is when many people will tell you the 'Paris of the North' should be seen. Several of our fellow travellers disembarked for some hectic nightlife, not to emerge from their cabins until we were cruising down the Risøyrenna Channel the following afternoon. Slender bridges connect the islands to the mainland and as the ship eased its way down narrow streams of sea-green water you could see the sandy bottom. Of the whole trip, this section was perhaps the most glorious, with mountains soaring up on either side and tiny clusters of houses nestling in the folds of the rock.

When we crossed the Arctic Circle we knew we only had two days left. Continuing down through spectacular fjords, we reached Kristiansund, an old shipping town founded by Christian IV of Denmark in 1641. It is dominated by the imposing Christiansholm Fort, a 17th-century fortress bristling with cannons, and the charming wooden houses of the Torget (market place). And so the journey was done. The following afternoon we ploughed through the smooth waters of the Sognefjorden and on to Bergen's attractive archipelago. 'The world's most beautiful voyage' had certainly lived up to its reputation and was one we will remember for the rest of our lives.

LEFT, *traditional clothing and footwear for sale in Hammerfest*

PRACTICAL INFORMATION

■ There are a number of ways travellers can use the Norwegian Coastal Voyage, also known as the Hurtigruten:

■ Package deals (including meals) offer 11-day trips from Bergen to Kirkenes and back (available all year).

■ Nine-day excursions to Spitsbergen that leave from Tromsø on the M/S *Nordstjernen* (June through August only). It is possible to combine this with the Norwegian Coastal Voyage.

■ A half voyage northbound or southbound, taking six days, starting in Bergen and flying home from Oslo via Kirkenes, or starting in Kirkenes and flying home from Bergen (available all year).

■ Cabins can be hired for 24-hour periods, which means that travellers can break their journey and do as much or as little of the voyage as they please.

■ Voyages can be booked through travel agencies or directly through the Hurtigruten – The Norwegian Coastal Voyage at OVDS, Sekretariatet, 8450 Stokmarknes (Tel: 76 15 14 22) or TFDS, Passasjeravdelingen, 9000 Tromsø (Tel: 77 64 82 00) or Bentours, Level 11, 2 Bridge Street, Sydney NSW (Tel: 2 9241 1353. Fax: 2 9251 1574.

■ The Hurtigruten is very busy in the summer months, so advance booking is recommended. Ships leave Bergen daily, and there are a number of shore excursions available, subject to demand. Warm clothing is essential.

Where the Majestic Shannon Flows

ROB STUART

'You're not thinking of taking that wee boat down the Shannon, are you?', called a fisherman from the bridge at Drumshanbo, a small village below Lough Allen on the upper reaches of the river. 'All the way to Limerick', I shouted back over the noise of the fast-flowing water. He was referring to my 10ft (3m) inflatable dinghy powered by a 5hp engine. He shook his head in consternation. 'There're snags of all kind down this stretch – rocks, concrete posts, wire, wrecked cars ... Knock your boat and engine to bits they will.' Nevertheless, with the help of a kindly cruiser and a bit of Irish luck, I made the 190-mile (304km) trip in four days and lived to tell the tale.

ABOVE, *pleasurecraft of all kinds ply the waterway network of the Shannon, which includes canals, loughs and tributaries*

BELOW, *the Shannon broadens as it flows gently past the town of Carrick*

In deference to the fisherman's warning, I used the oars instead of the engine until I reached deeper water at Acres Lough, around ½ mile (1km) downstream. Here, after battling through the obstacle course from Drumshanbo (the fisherman omitted to mention the almost impenetrable thickets of overhanging willows), I started up the engine and cruised gently through the lapping waters of this remote and tranquil lake. The surrounding countryside, though flat, was stunning. Most impressive of all was the sheer abundance of nature and mix of exuberant primary colours, from the peat-black water to the cerulean blue of the sky. As an old farmer I met outside Leitrim lyrically put it: 'When God flew over County Leitrim, so weighed down was he with nature, he dropped most of it here.' In addition – apart from the putter of the engine – there was the silence of the place, a quietness almost in reverence of itself. So frequently did I turn off the engine just to enjoy the peace, it took well over an hour to cross the lake.

ABOVE, *Carrick-on-Shannon, the county town of Leitrim, is now a popular cruising centre*

It is said that there is more myth, history and legend per mile on the Shannon than any other river in Great Britain; quite a lot to take in considering it is 214 miles (342km) long. In fact, since the re-opening of the Ballinamore and Ballyconnell Canal in 1995, linking the Shannon to Lough Erne in Northern Ireland, it provides, at 500 miles (800km), one of the longest navigation routes in Europe.

Drumshanbo, my starting point, must be one of the quietest villages in Ireland, a good retreat by any standards. Just downstream, Acres Lough is unbeatable for pastoral serenity.

On reaching the lively town of Carrick it is worth seeking out Ryan's Bar, especially for its excellent music nights, and the Carrick Craft and Emerald Star Line marinas are impressive.

After Jamestown there is a vast basin of loughs to be explored, the biggest being Tap, Boderg and Bofin. The area is a haven for wildlife and a good place to spot the elusive otter. Richmond Harbour is worth a visit, if only to wander the streets of colourful houses. It is certainly a pleasanter place than Lanesborough, blighted as it is by the twin towers of a local electricity generating station.

Once out on Lough Ree, the sheer size of this vast inland lake is breathtaking. Most interesting of its several islands is Inchcleraun (Quaker Island), once the home of Queen Maeve, the mythical pagan goddess of drunkenness.

Beyond Athlone is the ancient site of Clonmacnois – 'St Kieran's city fair' – which can be reached easily from the river.

To the south of Shannonbridge is Clonfert, in whose opulent 12th-century cathedral lies Europe's most famous travelling saint – Brendan the Navigator. Along this stretch you might just glimpse (or hear) the corncrake since the meadows near Clonmacnois are a favoured habitat of this now rare bird.

Lough Derg, the other large inland lake on the trip, also has a famous island – Inishcealtra (Holy Island) – which lies a mile (1.5km) off-shore near Mountshannon. Legend has it that St Colm landed there in the 6th century and discovered a tree which 'distilled juice tasting of honey and with the headiness of wine'. A mead tree, perhaps?

After passing by Ardnacrusha generating station, the river runs swiftly into Limerick, the last major town on the Shannon before it widens into its estuary and disappears forever into the Atlantic Ocean beyond.

ON TOWARDS CARRICK

I pressed on towards Battlebridge, with the intention of making it to Carrick-on-Shannon, around 25 miles (40km) downstream, before nightfall. The river now broadened out into soft, low-lying pasture, and there was barely a current in the water. On contemplating this sedate and rather beautiful waterway it was hard to imagine that the Shannon had once formed a bitter divide over which thousands of hapless Irish families had passed, banished from the rich agricultural land of Leinster to bleak poverty in Connacht by Cromwell in the 1650s. Carrick is a bustling, energetic town, and on the menu at Ryan's Bar, where I stayed the night, were salted-beef sand-

ABOVE, *Athlone's 13th-century fortress still stands guard beside the river's crossing point*

BELOW, *small-time business in Athlone*

wiches, Guinness and Seamus – a virtuoso fiddle-player. Did he know a song to celebrate my journey? He did, *The Shannon Waltz*. I sat transfixed as his fingers and bow glided effortlessly over his fiddle from which this deeply plaintive tune emerged. Later at the bar I asked him how much he knew of the Shannon. Seamus smiled: 'You can travel the Shannon for two lifetimes and still never know it properly.' To its compendium of history, legend and myth, add *mystery*, I thought.

The word was to be prophetic because the next day the Jamestown Canal, which provides a short-cut from Jamestown into Lough Tap, had apparently mysteriously disappeared. A couple of old gentlemen, sitting on a bench overlooking the river, seemed as baffled as I was when I called out to them for directions to the canal. 'Ballyconnell Canal?' one asked, with a puzzled look. Equally puzzled, the other chap shook his head. Meanwhile the boat and I were going round in circles. Finally, with a gleeful yell: 'Ach, you mean the Jimst'n Canal.' With this I was on my way. How could I have possibly mispronounced such a straightforward English name!

HITCHING A RIDE

There was one lock on the canal, kept by Michael, an amiable, soft-spoken man. He had bad news for me. 'Weather report's not good. You might just make it across Lough Boderg today, but Lough Ree's out of the question for the rest of the week, I reckon. Too rough, you see.'

Though deeply disappointed, I had no intention of abandoning the journey just yet. I asked him what I could do. 'Sparrow on the eagle's back', he grinned inscrutably. What did he mean? 'Hitch a lift on one of the pleasure cruisers. You should be able to pick one up at Roosky at the south end of Lough Bofin.'

Rough weather, turbulent water and an irregular shoreline that frequently took me off course made the crossing of Loughs Boderg and Bofin – both nearly 2 miles (3km) wide in parts – a bruising experience. However, a chance meeting with a launch crewed by three German fishermen provided a brief but welcome rest.

The Shannon is famous for its excellent pike fishing and there on the stern deck of the cruiser lay a freshly caught 15lb (33kg) fish. Were they going to eat it? 'Well, of course', said Peter. 'Cut up in steaks and fried in butter...' He pinched his fingers at his lips gesticulating gustatory satisfaction. 'Beautiful!' he exclaimed. To celebrate the catch we toasted the fish with large glasses of Powers whiskey.

At Roosky, I had reason to celebrate again. As Michael – the Jamestown Canal lock-keeper – had predicted, a large Derg-line cruiser was moored up there. After explaining to its crew – again Germans – that I needed a lift across Lough Ree the following day, they were only too pleased to help. 'Meet us at Lanesborough tomorrow morning at 8am sharp', said Wolf, a commercial lawyer from Berlin.

The sun was setting to an opposing blood-coloured moon, and though I had around 30 miles (48km) to travel before Lanesborough, like the river, I was in no hurry. Evening, I found, was the best time on the Shannon. As the light faded, the river in contrast seemed to brighten in the last rays of reflecting sunlight, defining it more sharply than I had ever seen it during the day. It was late autumn and the sky was full of birds mustering on the wing, getting ready to migrate.

Like so many of the small towns and villages along the Shannon's route, Lanesborough is made up of a broad main street flanked by tightly packed, pretty, if slightly jaded, terraced cottages, and with so many bars the short street can seem a long walk if you have a taste for Guinness. Equally intoxicating is the heavy, pervasive smell of peat-burning fires. After a delicious bowl of 'Shannon-size' soup, at Desiree's B&B where I stayed, I fell in with a late-night poker game at one of the bars. 'The Paddies will get you!' one of the players chortled. And they did. They emptied my pockets, but graciously left the shirt on my back.

As arranged, my lift across Lough Ree left promptly at 8am. The cruiser was the epitome of the floating gin palace, with the ice and lemon thrown in free. Sumptuous, comfortable and swift, I barely felt the 4ft (1m) waves we met during our four-hour crossing of this vast inland waterway. Measuring around 24 miles (38km) long, and in parts 5 miles (8km) wide, it rather gives the lie, along with the other loughs, to the claim that the Shannon is wholly a river. Edmund Spenser, the English Elizabethan poet, was correct to write of 'the spacious Shannon spreading like a sea'. Out in the middle of the lough, following the navigation route clearly indicated by large black buoys, landfalls were scarcely visible.

ABOVE, *Lough Ree, with the stone tower that is said to mark the very centre of Ireland*

THROUGH THE MIDLANDS

At Athlone, reputed to lie at the centre of Ireland, I set off for Portumna, by far the longest haul of the trip. This section of the river flows in broad meanders through countryside (called the 'Midlands') of exceptional beauty. In addition, like everywhere else on the Shannon, it is steeped in ancient history. Clearly visible along this stretch of the river is Clonmacnois, one of the

most remarkable of the early Irish monasteries. Founded by St Ciaran (or Kieran) in 545, it is the earliest recorded place of pilgrimage in Ireland. It was eventually ransacked by the English army in 1552. Unfortunately, St Ciaran chose as the site for his monastery what was to become over successive centuries a place of considerable strategic and commercial importance: the *Eiscir Riada* – the most important east–west roadway in ancient Ireland. An old ford, evocatively named 'At Swim Two Birds', where Ireland's most famous navigator, St Patrick, once crossed the river, but whose exact location is a matter of conjecture, is meant to mark the crossing-point.

To save time, I decided to go over the weir at Meelick rather than wait my turn to go through the lock. Misjudging the strength of the current, however, the boat went over before I had time to lift the engine fully out of the water. The propeller caught the weir's concrete parapet and the sheerpin broke. Fortunately I had several spares with me. However, this incident and a later oversight whereby I ran out of petrol and had to trudge for what seemed like miles through bog and woodland to find a farm that could spare me some fuel, put me behind schedule. In fact, I didn't arrive at Portumna until well after dark but was compensated by a sky heavy with stars and moonlight so bright that the river shone like an illuminated runway.

Rough Water

The small town of Portumna is situated at the head of Lough Derg, another vast inland lake. This time I had no offer of a lift but on venturing on to the lough the next morning the water was deceptively calm. It wasn't long, however, before those 'mutinous waves', as Irish writer James Joyce called them, started battering at my boat. Again, as on Lough Ree, the landfalls virtually disappeared as I motored out from Terryglass where I briefly stopped to consult a fellow mariner's Admiralty maps of the lough. George, the owner of an old No 71M Shannon barge, had some advice for me: 'Keep out in the middle. That's where the route is. You'll see the buoys there.' Out in the middle turned out to be a tumultuous sea. I decided, eventually, to pull in at Williamstown marina, around halfway down the lough. John, the manager, greeted me warmly. 'You were wise to pull in here', he said. 'Off Parker's Point, a bit further down, the wind would really catch you.'

John kindly arranged a lift for me – on this occasion, a 'sparrow on a *taxi's* back' – down to Killaloe, the most southerly point on the river for large pleasure craft navigation. Charlie, my taxi-driver, was a typical Irish raconteur. For a solid half-hour he talked about his past – how he had

ABOVE, *founded by St Ciaran, the monastic site of Clonmacnois became a centre of art and learning*

RIGHT, *the Romanesque doorway of Clonfert Cathedral, south of Shannonbridge*

worked in a paint factory in New Zealand and travelled across Australia and shaken hands with Bobby Kennedy. He was in his 70s and courting a lady from Edinburgh.

Once back on the river I motored off on the last leg of the journey to Limerick, around 32 miles (51km) downstream. Beyond O'Brien's Bridge I entered the 8-mile (13km) headrace, a section of canal leading to the impressive Ardnacrusha hydro-electric station. Built between 1925 and 1929, it was the most important building project in the early history of the state and brought Ireland into the 20th century. It had a surprise in store for me. Once in the lock, I was dropped 150ft (45m), and I reckoned something in the region of 1.5 million gallons (6.75 million litres) of water were displaced for my 'wee' boat.

Renowned as a gently flowing river, the Shannon's character changes dramatically at Limerick with the rapids there. Rather than mooring up beforehand, I decided to tackle them and it was the only occasion on my four-day trip when I came perilously close to being capsized. Mercifully I survived – guided, I like to think, as much by the kindly spirit of the Lady Sinann, after whom the Shannon is named, as by my own efforts.

LEFT, *Lough Derg, surrounded by desolate moorland and bog*

Practical Information

■ While much of the river is safe for small boats, extreme care should be taken when boating on the larger loughs (lakes). Wear a life-belt at all times and take suitable wet-weather clothing. Also, consult local people about water and weather conditions.

■ The cruising season is between March and October. Between late October and early March the locks have restricted operational times. Consult cruising companies.

■ Cruisers can be rented from: Athlone Cruisers Ltd. Tel: 0902 72892. Fax: 0902 74386.
Ballykeeran Cruisers Ltd. Tel: 0902 85163. Fax: 0902 85431.
Carrick Craft (bookings UK & International). Tel: (44) 1734 422975. Fax: (44) 1734 451473.
Derg Line Cruisers. Tel: 061 376364. Fax: 061 376205.
Emerald Star Line Ltd. Tel: 01 679 8166. Fax: 01 679 8165.
Shannon Castle Line. Tel: 01 660 0964/01 660 0588. Fax: 01 668 9091.
Silver Line Cruisers. Tel: 0509 51112. Fax: 0509 51632.
Lowtown Cruisers Ltd. Tel: 845 60532. Fax: 845 60532.
(All are members of IBRA – the Irish Boat Rental Association.)
Club Seafarer, New Zealand (also covers Australia). Tel: 9 579 8097. Fax: 9 525 2281.

■ Further information on cruising the Shannon can be obtained from the Irish Tourist Board:
London: Tel: 0171 493 3201. Fax: 0171 493 9065.
Dublin: Tel: 01 602 4000. Fax: 01 602 4100.
Sydney, NSW: Tel: 2 9299 6177

ABOVE, *gannets, found all around the coast of Britain, can often be seen diving for fish in spectacular fashion*

Britain's Island Odyssey

ELIZABETH CRUWYS AND BEAU RIFFENBURGH

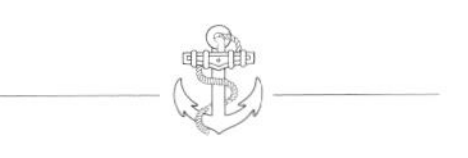

The Island Odyssey takes passengers on a tour of some of the most inaccessible places in the United Kingdom. From the gentle woodlands of Devon, to remote islands that are teeming with birds, *Caledonian Star* weaves under towering cliffs, past lines of golden sand dunes and into marshy estuaries. This beautiful journey begins in England, heads south to the Channel Islands, then west to the Isles of Scilly and Ireland, and north to the Hebrides, Orkney and the Shetland Islands, finally turning south to Fair Isle and ending in Scotland's ancient capital, Edinburgh. There is something for everyone, from medieval castles and abbeys and gorgeous gardens exploding with brilliantly hued flowers to wild sea-swept rocks and bird-filled skies.

On an afternoon in late spring, the train pulled into Dartmouth railway station and we disembarked, armed with binoculars, wet-weather gear, cameras and an entire library of bird-watching books. Struggling with our heavy luggage in the bright May sunshine, we began to wonder whether we had been overly pessimistic in our predictions about the weather, and whether two pairs of boots and multiple layers of thermals and fleeces were really necessary. But few places have reputations for rain and wind like Scotland and Ireland and so, valiantly, we hauled our mammoth suitcases to the quay, where *Caledonian Star* was waiting. Not dumping the extra gear there and then was a decision for which we were later very grateful.

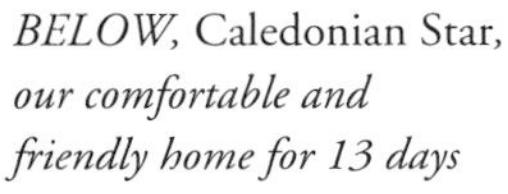

BELOW, Caledonian Star, *our comfortable and friendly home for 13 days*

Caledonian Star is ideal for a tour of Britain's islands. She is large enough to be able to make good time between stops, but small enough to be able to nose her way into spots inaccessible to bigger ships. All 68 of her passenger cabins face outside, which means that those not wanting to brave the elements – and, at times, only the very brave did – miss nothing by remaining indoors, even in bed! With a full complement of 110 passengers and an informal dining room, it was possible to get to know fellow travellers. Many of them had done their research well, and there was always someone on hand willing to tell us about the birds and the history of the places we visited, providing an entertaining addition to the on-board guest lecturers.

ABOVE, *handsome houses line the waterfront in the historic naval town of Dartmouth*

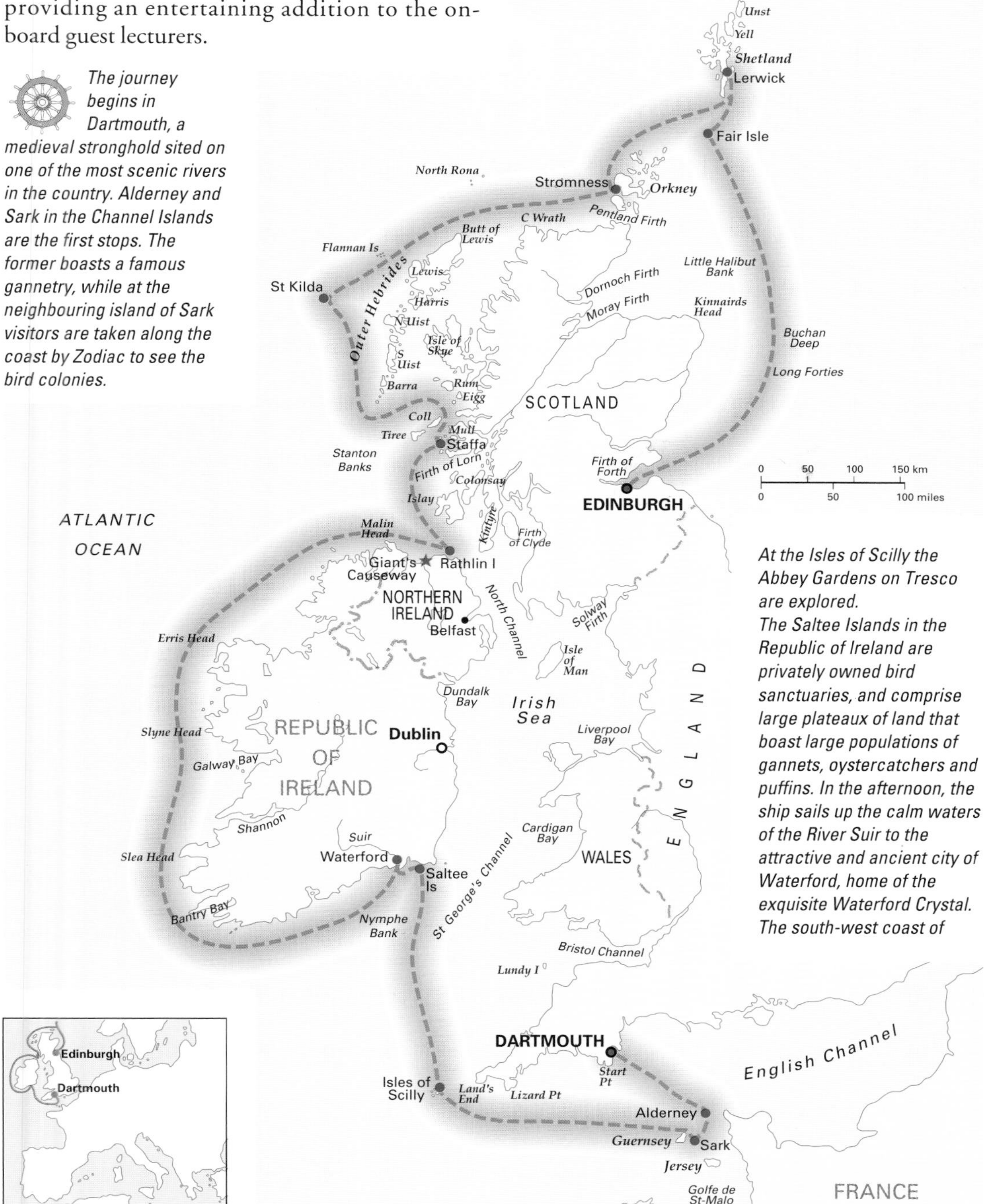

The journey begins in Dartmouth, a medieval stronghold sited on one of the most scenic rivers in the country. Alderney and Sark in the Channel Islands are the first stops. The former boasts a famous gannetry, while at the neighbouring island of Sark visitors are taken along the coast by Zodiac to see the bird colonies.

At the Isles of Scilly the Abbey Gardens on Tresco are explored.
The Saltee Islands in the Republic of Ireland are privately owned bird sanctuaries, and comprise large plateaux of land that boast large populations of gannets, oystercatchers and puffins. In the afternoon, the ship sails up the calm waters of the River Suir to the attractive and ancient city of Waterford, home of the exquisite Waterford Crystal.
The south-west coast of Ireland, with rugged cliffs plunging into heaving seas, provides some spectacular sights, including Great Skellig, the haunt of some 18,000 gannets. After the gannetries, Caledonian Star *continues north, round the coast of Sligo and Donegal to Rathin Island with its bird stacks, and past the Giant's Causeway.*
The Outer Hebrides present a different seascape, comprising low, windswept islands with long beaches and golden lines of sand dunes. St Kilda is one of the most remote of Britain's islands, located 110 miles (175km) out into the Atlantic. Zodiacs land passengers at ancient Hirta, an abandoned village now in the care of the National Trust for Scotland.
At the Orkneys, Caledonian Star *puts in at Stromness where transport is available to take passengers to see Skara Brae, a neolithic site, and past Scapa Flow, where the German fleet was scuttled in 1919.*
The Shetland Islands offer more wild and rugged coastline, as well as Lerwick, an elegant city with a 7th-century fortified tower and a Viking exhibition.
The last port of call is Fair Isle, another remote island where birds outnumber the few remaining islanders by the thousand, before the ship finally arrives in Leith Harbour in Edinburgh.

RIGHT, *horse-drawn carriages and tractors are the only form of transport on the little island of Sark*

Dartmouth is a charming town, hugging the banks of the mellow River Dart. Its castle, perched on a rocky promontory that juts into the sea at the mouth of the river, was begun in the 14th century but left unfinished. In the 15th century, after repeated raids from French pirates, residents became so exasperated that they took matters into their own hands and completed the castle themselves. A second castle was built on the opposite side of the river at Kingswear at the same time, so that any French pirates foolish enough to fix their greedy eyes on the prosperous little town would have to run a formidable gauntlet of cannon fire before they could reach it. We faced no such barrage and, in the golden sunlight of late afternoon, slipped quietly down the river and out into the open sea.

South Coast Islands

Immediately upon reaching the sea, the wind began to blow and on went the first layer of thermal clothes. The second layer followed as the sun set, scattering multi-coloured lights across the sea and disappearing in a ball of flaming red. Gulls had followed us out of the harbour and they continued to pursue us as night fell, small flashes of white standing out against the dark sea.

BELOW, *appropriately named Gannet Rock, off the coast of Alderney*

The following morning we arrived at out first port of call, Alderney, in the Channel Islands. St Anne is the capital, a charming little town with something of a French air and narrow, winding streets and brightly coloured buildings. It was peaceful, too, the silence broken only by footsteps clattering on the cobbles and the mewling of gulls. In the afternoon we enjoyed a pleasant cruise along the cliffs by Zodiac before landing on Sark. The harbour nestles at the foot of steep-sided hills and transport to the top of the cliffs is via a trailer pulled by a tractor – there are no cars on Sark. A horse-drawn cart then ferried us at a leisurely pace along narrow lanes bordered with hedges and littered with a dazzling array of wildflowers, from the creamy white of cow parsley and daisies to the delicate pinks and reds of campion and pimpernel, and from the vivid yellows and golds of buttercups and vetch to the blues of forget-me-nots and speedwell.

Sark is still a feudal fief, the centre of which is the 16th-century house of La Seigneurie with its amazing walled gardens. The walls were raised from the 1830s onwards and protect a huge number of subtropical plants, all thriving in the mild climate. On the way back to the ship, the Zodiacs nosed in and out of the rocks and we were able to lean over the side and see fish feeding among waving fronds of kelp. One species of seaweed is called 'japweed' (*Sargassum muticum*), which, legend has it, was introduced into British waters on the propellers and keels of Japanese ships.

The next day we were up early to catch the first glimpse of the Scillies, a windswept collection of 150 islets and skerries located about 28 miles (45km) south-west of Lands End in Cornwall. As we drew nearer the islands gradually emerged out of the mist, the white sandy beaches and turquoise seas looking almost Caribbean. The island of Tresco not only boasts Cromwell's Castle, which dates from the mid-16th century, but the fabulous Tresco Abbey Gardens. They were founded in the 1830s by the first Lord Proprietor of the Isles, Augustus Smith, who planted a 'shelter belt' of saplings on an island that was hitherto treeless. His great-nephew later made plant-foraging expeditions to the southern hemisphere to collect specimens to add to the collection.

A trip to the island of St Agnes allowed us to stretch our legs by walking to the lighthouse at the far end of the island – the second oldest lighthouse in the country, dating from 1680 – where we saw scores of brightly coloured butterflies dancing across the sand dunes.

Ireland's Coastline

The following morning saw us at Great Saltee, a vast platform of rock tilting out of the sea off the south-east coast of Ireland which is home to thousands of seabirds, including puffins, kittiwakes, gannets, guillemots, razorbills, cormorants, oystercatchers and several species of gull. The cliffs at the southern end are a sight that almost defies description, where birds circle, wheel, dive and glide in vast numbers, and the sea is littered white, black and grey as they skim and wallow on the surface. The noise, too, is incredible, with caws, screeches, and cries battling for dominance over the roar of the sea. In places, the cliffs are stained white with guano.

Waterford is a jewel among Irish cities. Big, noble-looking herons watched us sail past up the River Suir so that little seemed to have changed since the first Viking settlers paddled upstream in the 8th century. Waterford is perhaps best known for its famous crystal, an industry that was founded by two brothers in the 18th century. The tour of the town included a visit to Mount Congreve, where there is a Georgian house surrounded by one of the finest gardens in the country.

LEFT, *the Giant's Causeway in Northern Ireland: one of the many legends surrounding this strange rock formation is that it was built by a giant, Finn MacCool, so that he could step across to Scotland*

Bottlenose dolphins and seals frequent the stormy waters around Glengarriff, south-west of Cork, where there were yet more gannetries, and the Zodiacs allowed us to inspect mammals and birds at quite close quarters.

The next stop was at Rathin Island in Northern Ireland, famous for its bird stacks, and we also saw the basaltic Giant's Causeway, as impressive from the sea as it is from the land.

North to Scottish Waters

The little island of Staffa in the Hebrides was made famous by Mendelssohn's overture *Fingal's Cave*, which we were able to enter by Zodiac. The fact that it was at the crack of dawn did much to enhance the ethereal cathedral-like atmosphere of this magnificent geological formation. Great

BELOW, *view of Glengarriff nestling at the head of an inlet of Bantry Bay*

ABOVE, *the village of Skara Brae, inhabited some 5,000 years ago, is remarkably well preserved*

columns of basalt soar upwards, and the booming and crashing of the sea against the rocks set up strange echoes in the cavern so that it was easy to understand how the composer was so inspired. Meanwhile, away from the buffeting of the sea, is Iona – selected by St Columba as a refuge from the world in the 6th century.

St Kilda is the stuff of legends, a cluster of three tiny islets lying some 110 miles (175km) off the coast of mainland Scotland. Owned by the National Trust for Scotland, the islands have a great deal to offer, ranging from the village of Hirta, abandoned in 1930 and undergoing restoration, to the spectacular St Kilda Cliffs. These vast granite faces plunge sheer down into the waves. One of the cliffs is the mighty Atlantic Wall of Conachair, an intimidating structure strewn with ropes left by climbers. In terms of birds, St Kilda is remarkable. It boasts the largest population of gannets in the world (some 50,000 pairs), together with what is reputed to be Britain's oldest and largest fulmar colony (63,000 pairs). When all the other species are added in, St Kilda's bird population comes to a staggering 400,000 pairs, making it one of the premier congregations of birds in the North Atlantic. It has to be said that it certainly smelled like it!

Some 6 miles (10km) off the north coast of Scotland lie the Orkneys, a group of 67 magical islands whose history stretches back to neolithic times. Maes Howe is a chambered tomb covered by a mound 26ft (8m) high and 115ft (35m) in diameter; it is oriented so that the setting sun of the winter solstice shines down the passage to the back wall. When the Victorians broke into it in 1861 they found that the Vikings had been there before them: although some of the walls were adorned with runic graffiti, the tomb contained nothing else. It is thought that 12th-century marauders made off with the gold and silver

PRACTICAL INFORMATION

- A number of cruise companies take passengers around Britain, although this trip was run by Noble Caledonia, 11 Charles Street, London W1X 8LE. Tel: 0171 409 0376. Fax: 0171 409 0834. In Australia contact Adventure Associates, 197 Oxford Street, Bondi Junction NSW 2022. Tel: 2 9389 7466

- The Island Odyssey takes 13 days in late May. Itineraries depend on local weather conditions.

- A similar trip, concentrating on the islands' gardens is also available in June and early July. Both trips only run once a year, so early booking is advisable.

- Landings are by Zodiac in some places, and wet-weather gear and wellington boots are essential. Anticipate spending a large amount of time outside and take plenty of warm clothing.

hoards that were believed to have been buried with the 9th-century king.

Stromness is a ferry port with a village atmosphere, its 18th- and 19th-century houses clustered around narrow streets. Eight miles (13km) to the north, overlooking the Bay of Skaill, is Skara Brae, a remarkable village that was preserved for 5,000 years under a sand dune. It lay completely hidden until a particularly ferocious storm blew away the sand in 1850.

More ancient history was in store for us on the Shetland Islands, another rugged group of about 100 islets located about 60 miles (100km) to the north of Orkney. These remote islands were under Viking rule until 1469 and are almost as much Scandinavian in flavour as they are Scottish. Isolated crofts and tiny fishing villages litter the coastline, although only about 15 of the islands are inhabited; the rest are the domain of birds and seals. Puffins (called 'Tammie Nories' in the Shetlands) and diving skuas jostle for space with the much rarer Pallas' sandgrouse and red-necked phalarope. We felt honoured when we glimpsed a minke whale in the water, while grey seals with their distinctive 'Roman' noses watched us from the rocky shores.

EDINBURGH BOUND

Heading back south, we reached the aptly named Fair Isle. As we circumnavigated it we were watched by thousands of wheeling seabirds, some of which dipped and dived in the wake of the ship. Fair Isle is noted for its storms and, as we approached, the sea heaved with a substantial swell, making a landing difficult. We contented ourselves with views from the ship until the weather relented and we were able to make a foray ashore for tea and scones in the village hall.

Our journey was coming to an end and we turned southwards for the final run to Edinburgh, stopping at Incholm Island Abbey on the way. We disembarked at Leith and were taken to Scotland's ancient capital. What better way to end a tour of some of Britain's wildest coastal scenery than with one of its most enduring castles? The castle site was certainly used as a residence by the Scottish king Malcolm Canmore nearly 1,000 years ago, although it had served as a secure refuge for many earlier inhabitants. Peering over its walls down the precipitous cliffs, we felt a certain affinity with the millions of birds we had watched – also safe in their rocky eyries hundreds of feet high.

LEFT, *Edinburgh Castle, dating in part from the 12th century, dominates the city from its great volcanic outcrop*

OPPOSITE PAGE, *Stromness has a long and illustrious history as Orkney's principal harbour town, a role it still holds today*

Cruising the Romantic Rhine

ELIZABETH CRUWYS AND BEAU RIFFENBURGH

ABOVE, *colour in Cologne's Altstadt, the 'old town' lying south of the cathedral*

BELOW, *good views of Cologne's famous twin-spired cathedral can be had from the river*

The River Rhine (Rhein) is a great ribbon of water than snakes through western Europe, having carved out some of the most spectacular scenery in the region. Steep-sided cliffs thick with trees plunge straight downwards, while castles cling precariously to rocky outcrops high above them, or stand isolated on tiny islands in the middle of the river. The journey from Mannheim to Cologne (Köln) is the Rhine at its best, featuring magnificent cathedrals, vast, undulating vineyards that produce some of Germany's best-loved wines, gorgeous castles and splendid views.

One of the words most frequently associated with the River Rhine (Rhein) in popular literature is 'romantic', and even a short trip along Europe's most important waterway justifies that description. From its source in the Swiss mountains to its mouth at the North Sea, the Rhine is a river where superlatives can be used without exaggeration. An international waterway since 1815, it is some 865 miles (1,390km) long and winds through six countries – Switzerland, Liechtenstein, Austria, Germany, France and Holland. Between Lake Konstanz and Rotterdam it drops about 1,310ft (398m), passing castles, cathedrals, palaces, huge modern industrial plants, vineyards, mountains, dykes, cliffs, forests and windmills.

To enjoy all the Rhine has to offer would take weeks, perhaps even months, but we had just a few days and decided to concentrate our journey in the Middle Rhine (Mittelrhein), starting in the Rheingau wine-growing region at Mannheim and ending in the Lower Rhine (Niederrhein) at the handsome cathedral city of Cologne (Köln), a distance of only 102 miles (163km).

For those in a hurry, comfortable floating hotels slice along this section of the river in just a few hours, but for those with more time at their disposal it is infinitely more rewarding to jump on and off the many ferries as the fancy takes you and explore some of the less well-known corners of this fascinating region.

The Rhine (Rhein) is one of the busiest rivers in the world, not only revealing its splendours to millions of visitors but providing an important cargo transport route. Inundated with household waste, fertiliser run-off from farms and vineyards, and effluent from factories, the Rhine is struggling to survive. Several species of fish and plant no longer live in it, and it is considered to be one of the major pollutants of the North Sea. While the 1980s and 1990s have seen considerable public pressure for water authorities to impose stricter measures, there is still a long way to go before this beautiful river can revert to its untainted state.
It is often said that no city except Basle truly spans the Rhine – it is simply too wide. Mannheim and Ludwigshafen face each other across 920ft (280m) of river, but, despite their proximity, are two completely separate cities. Mannheim, first recorded in the 8th century, is one of the largest inland ports in Europe. Ludwigshafen, which developed as a suburb of Mannheim, is dominated by the enormous BASF factory.
Mainz and Wiesbaden also face each other, with Wiesbaden occupying the wooded slopes of the Taunus Hills. It was settled by the Romans, who also built a fortress at Mainz, and a bridge between the two has existed since the 1st century AD.
Bingen is a gorgeous old town with some splendid baroque architecture, while on the east bank stands Rüdesheim, famous for its wines. The Drosselgasse is a lively street lined with taverns in which local produce can be sampled, while the Niederburg, once held by the archbishops of Mainz, is now a museum of wine.
The area is packed with lovely old castles, including Mäuseturm on its rocky island, Reichenstein (first raised in the 11th century) and 13th-century Rheinstein. Loreley, the fabled home of the siren who lured sailors to destruction, towers above the river.
Magnificent Rheinfels, dating from 1242, was once the mightiest fortress on the Middle Rhine, while near by, Marksburg is a vast 13th-century stronghold.
Koblenz, at the confluence of the Rhine and Mosel, has long been a site of strategic importance and boasts the Electoral Palace, the former residence of the Electors of Trier and a Romanesque church.
Remagen Bridge was the scene of fighting in World War II. It collapsed in 1945 and one of its ruined towers has been preserved as a museum of peace.
Bonn, the seat of government for the Federal Republic of Germany from 1949 to 1991, not only has a number of fascinating buildings with political connections, but is also the birthplace of Ludwig van Beethoven.
Cologne suffered enormously from allied bombing during World War II but its fabulous cathedral escaped serious damage and is one of the best Gothic buildings in Europe.

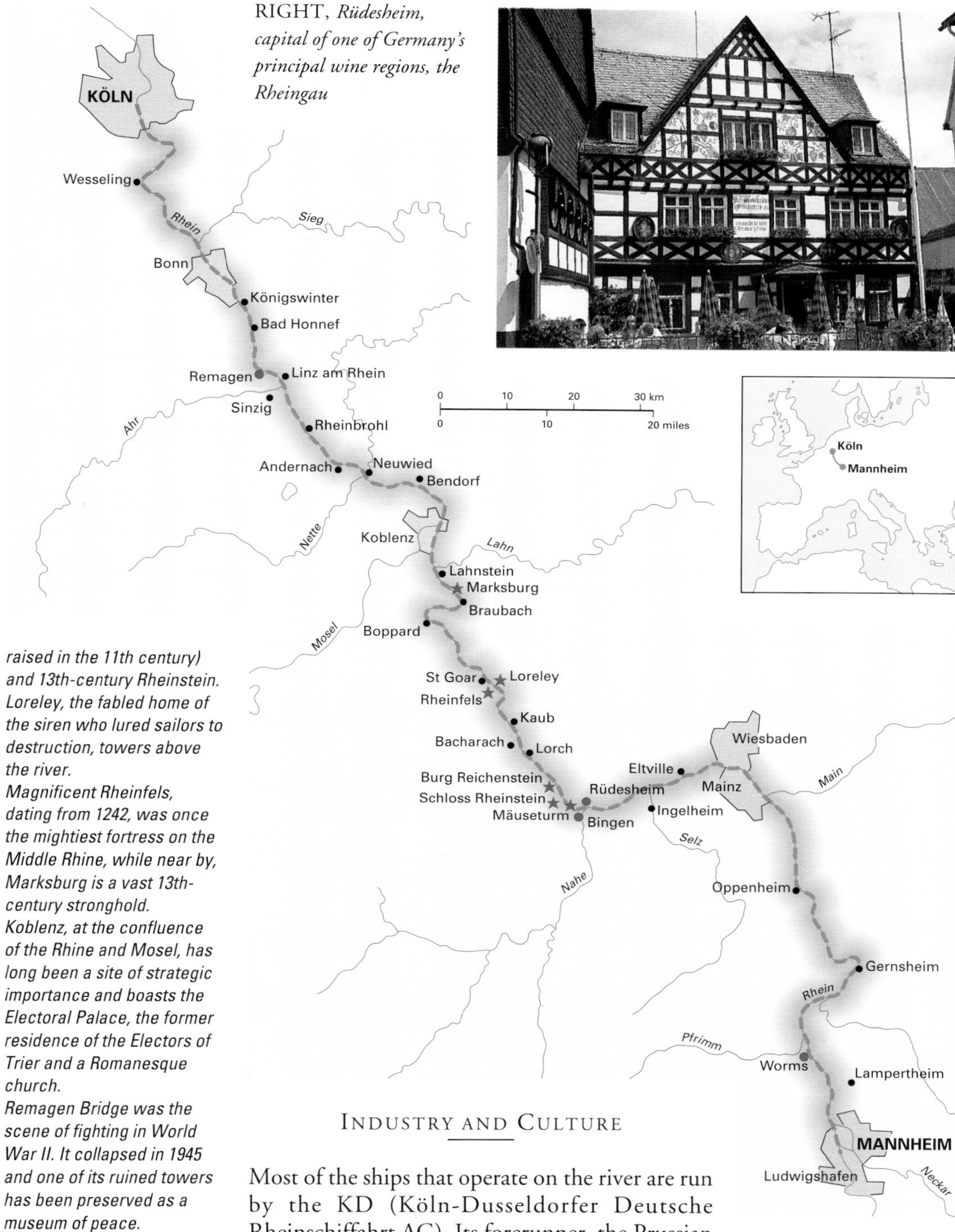

RIGHT, *Rüdesheim, capital of one of Germany's principal wine regions, the Rheingau*

INDUSTRY AND CULTURE

Most of the ships that operate on the river are run by the KD (Köln-Dusseldorfer Deutsche Rheinschiffahrt AG). Its forerunner, the Prussian Rhine Steamship Company, began a regular passenger service between Mainz and Cologne in the early 20th century and by 1913 it was carrying 2 million passengers per year. Today, the fleet comprises eight cruise ships and a variety of motor ships, paddle-steamers and hydrofoils. If you travel with KD, you will be in exclusive company: distinguished passengers have included Queen Victoria (1845), Prime Minister Nehru (1956), President de Gaulle (1961), Emperor Hirohito (1971) and President Ford (1975). It goes without saying that the KD fleet boasts luxurious cabins, impeccable service and excellent food.

We saw some magnificent sites before we even boarded ship in Mannheim. Our trip actually began across the river, in the city of Ludwigshafen. It was named in 1843 after Ludwig I of Bavaria, although a fortress had been built there as early as the 17th century. Ludwigshafen's most impressive feature is the vast BASF (Badische Anilin- und Sodafabrik) plant that is locally proclaimed as the largest single industrial site in Europe. It straggles for miles along the banks of the Rhine – a huge city of pipes, chimneys and warehouses. We were invited on a guided tour (conducted in German) and the sheer size of the site, which is involved in computer services, magnetics and the production of industrial and fine chemicals and reactive resins, was mind-boggling.

The day before, we had taken a trip some 12 miles (19km) out of Mannheim, along the River Neckar, to the ancient university town of Heidelberg. The university was founded in 1386 and the town boasts a number of medieval buildings in addition to the Schloß, built of red sandstone and formerly one of the finest examples of German Renaissance architecture. Although ruinous today, having been virtually destroyed by the French in 1689, the castle remains a magnificent sight.

BELOW, *the attractive university city of Heidelberg and its romantic castle lie on the banks of the Neckar, a tributary of the Rhine*

The ship docked near the Schloß in Mannheim, a splendid ducal palace dating from the 18th century. Despite its antiquity – there was a settlement here in the 8th century – the town is organised on a grid pattern and it is almost impossible to get lost. A major port, Mannheim is also famous for the first foot-propelled bicycle (1817) and for the motor vehicle demonstrated by Carl Benz in 1886.

Early in the morning the ship nosed off down the river, passing a number of great vineyards before reaching the ancient town of Worms. The glorious Cathedral of St Peter was consecrated in 1018, while the synagogue in Judengasse is the oldest in Europe. Almost completely destroyed during Kristallnacht in 1938, it has been well restored and is open to the public.

Mainz is another ancient university town, boasting a six-towered cathedral that was first founded in the 10th century, although most of the present building is later. The city is perhaps best known as the birthplace of Johannes Gutenberg (about 1397), the man generally credited with inventing the printing press. A Gutenberg Bible is on display in the charming museum which also gives an informative guide to the history of the book since the 5th century.

On the eastern bank of the Rhine stands the equally ancient city of Wiesbaden, where the Romans built a temple to Jupiter and took advantage of the hot springs in the 2nd century BC. The spa waters still gush forth at a temperature as high as 66°C (150°F) from a depth of about 6,560ft (1,994m). Wagner composed *Die Meistersinger von Nürnberg* here, while Brahms found the inspiration for his Third Symphony. The city has also been visited by Goethe, Schumann, Turgenev and Dostoevsky, although apparently the latter two came for the gambling casinos rather than the music.

VINEYARDS AND FORTRESSES

The whole of this section of the Rhine is noted for its white wine and the undulating, sun-washed slate slopes of the Rhenish Uplands produce some of the best white wines in the country, including Rheingau Riesling. Towns such as Eltville, Erbach, Geisenheim, Hattenheim, Hallgarten, Kiedrich, Mittelheim and Rüdesheim are all noted for their high-quality vineyards and provide a compact route for those travellers interested in sampling some of the local offerings. Rüdesheim has a museum of wine, taking visitors back through more than 2,000 years of viticulture. The town has been strategically important since

medieval times and was protected by no fewer than four castles.

On the west bank stands Bingen, guarding the Bingen Loch where the river cuts a narrow passage through the hills. Near by is the Mäuseturm, an early medieval toll tower standing on a rocky islet. Legend has it that greedy Bishop Hatto was eaten by mice in the tower as punishment for starving his parishioners. Bingen was also the home of Hildegaard (1098–1179), a remarkable Benedictine nun who was poet, playwright, philosopher and musician. We heard a concert that included some of her music in the Parish Church of St Martin, where the pure voices of a women's choir echoed evocatively around the soaring Gothic arches.

The next day, on leaving Bingen, we were overawed by a whole flurry of castles and fortresses, many of them perched high above the river as it heads directly north. Reichenstein, first built in the 11th century, with the reputation of being the stronghold of a robber-baron, has been destroyed and rebuilt many times. Rheinstein formerly served as a toll station for the Archbishops of Mainz in the 13th century and perches some 260ft (79m) high, on a rocky crag. It was restored in the 19th century by a Prussian prince. Opposite stand the ancient ruins of Ehrenfels, clinging precariously to a hillside and surrounded by vineyards. It was built in 1211 by the Archbishop of Mainz, but was a ruin by the 17th century. Next come handsome 11th-century Schöneck, splendid 10th-century Schönburg (now a hotel), and the curious Pfalzgrafenstein, a little 14th-century fortress standing in the middle of the river like a warship.

Of course, there is far more to the Rhine than castles, and even those of us who have a special love for old fortresses were pleased to see one of the best known geological features of this part of the river – Loreley. This slate crag towering some 435ft (132m) above the river is the fabled home of the nymph who lured boatmen to destruction by her singing, although on a bright summer afternoon all we could hear was the constant chug of diesel engines.

We passed two more magnificent castles before reaching Koblenz. Rheinfels, perched on a site

ABOVE, *Burg Katz (Cat Castle), 14th-century fortress of the Counts of Katzenelnbogen, with the fabled crag known as Loreley beyond*

ABOVE, *built at the meeting point of the Rhine and the Mosel, Koblenz marks the midpoint of the cruise*

375ft (114m) above the river, was once the largest fortress on the Middle Rhine and dates from 1242. Marksburg is thought to be the only hilltop fortress on the Rhine to have escaped destruction during the 17th century; a vast 13th-century stronghold, it now contains holdings of armour and armaments.

Our day ended at Koblenz, which stands at the confluence of the Rhine and the Mosel at a site fortified since Roman times. This small city is also a bustling port, boasting some splendid 18th-century palaces, a 13th-century castle and a Romanesque church founded in the 8th century and dedicated to St Kastor. Each August, the area between Koblenz and Braubach on the opposite side of the river hosts the 'Rhine in Flames' festival, featuring its impressive firework display.

We knew next to nothing about Linz before we arrived there the next day but it turned out to be a highlight of the trip, a pretty little town with exquisite half-timbered buildings and the scanty remains of medieval fortifications.

We knew more about our next stop, Remagen Bridge, however, as did most of the other travellers with any interest in World War II. Built between 1916 and 1918 on the orders of the brilliant General Erich von Ludendorff at a strategically important crossing point, it was taken by US troops in 1945 and played a vital role in the subsequent invasion of Germany. The bridge collapsed ten days after it was seized and the tower on the left bank has been preserved as a monument to peace.

ANCIENT CITIES

Soon we reached Bonn, the old capital city of the Federal Republic of Germany. Despite its lost status, Bonn is a charming place with a lively market place in which we spent several hours just enjoying the flavour of the area – as well as the flavour of the local wines on sale in the taverns! The 18th-century Rathaus dominates the square, while near by is St Remigius' Church and the Romanesque Münster. For us, the chief charm of the city was its association with Beethoven, including the house in which he was born, now a museum. It features original manuscripts and one of his organs, plus a variety of instruments through which the composer tried to overcome his incipient deafness.

We did not have to travel far on the final day of our journey, ending at the busy cathedral city of Cologne. Many people know the story of the cathedral and how it miraculously escaped the hail of bombs dropped by the allies during World War II. It can be seen from the river, a massive, timeless structure with twin towers shooting up into the sky and dominating the skyline. Building work started in 1248 but like many medieval structures it grew in stages and was not completed until the 19th century.

Cologne has much else keep the traveller occupied – there are museums, churches, medieval town gates, and even the remains of a Roman fort. Not to be missed is 4711 Glockenstraße, where the famous 'eau de Cologne' was made.

After Cologne, the KD ships continue north into Holland where the Rhine meanders less urgently through the flat countryside of northern Europe with its dykes and windmills. But our journey was done and, with regret, we turned our backs on the mighty river and its chemical plants, castles, vineyards, churches and docks and made for the railway station, conveniently located next to the cathedral.

PRACTICAL INFORMATION

■ Transport on this part of the Rhine is dominated by the company KD (Köln-Dusseldorfer Deutsche Rheinschiffahrt AG). KD has offices all over the world, contacts for which can be obtained from German tourist bureaux. The main German office is at KD Deutsche Flusskreuz 221 2088217.

■ Travellers can journey along the Rhine in two different ways. They can book an all-inclusive package holiday on one of the cruise ships for periods ranging from four to 15 days, in which all food, accommodation and shore excursions are included. Alternatively, they can travel short distances on the excursion ships, and arrange their own accommodation along the way.

■ The Rhine area can become very busy during the summer months, and early booking is recommended.

LEFT, *Bonn's finest church, the Münster, with Beethoven's birthplace (inset), now a museum*

FAR LEFT, *the impregnable Marksburg has stood intact for nearly 800 years*

Sailing the Med Clipper Style

PETER AND HELEN FAIRLEY

As a boy, Mikael Krafft lived next to a shipyard in Sweden, where the old-timers told him tales of the great sailing ships which once dominated the high seas. Krafft dreamed of building one of his own – a modern, hi-tech re-creation of classic clipper ships such as the *Cutty Sark*. In fact he has built two. His company, Star Clippers, operates one in the Caribbean and the other in the Mediterranean. We boarded the *Star Flyer* for one of its regular relocation cruises – from the western Mediterranean to the Greek Islands – and found luxury under the sails.

ABOVE, Star Flyer's *traditional rigging*

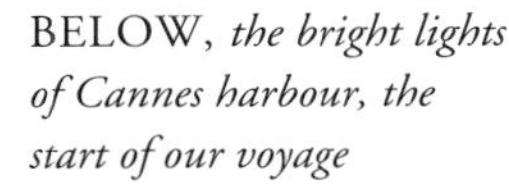

BELOW, *the bright lights of Cannes harbour, the start of our voyage*

'You are not coming on a cruise – you are beginning a voyage.' With those first words from the captain, my wife and I realised that we were in for something 'different'. We had just boarded the *Star Flyer* by one of its orange lifeboats, which double up as tenders when the schooner is in port. Somehow, our bags were already waiting on deck, along with a mass of other luggage belonging to the 170 other passengers from seven different countries. 'Don't worry – it'll be taken to your cabin' said the purser, whose neatly pressed white blouse and shorts, ankle socks and tennis shoes would have graced the Centre Court at Wimbledon. 'Come and have a rum punch and enjoy the moonlight.'

The scene was romantic by any standards. Cannes harbour was a mass of twinkling lights. The *Star Flyer*, dressed overall from her 360ft long (109m), white steel hull to the tops of her four masts, was clearly the star attraction, to judge by the little boats which clustered around as the crew first raised, then braced our 16 sails.

To reach our cabin we had to pass through the dining room, where a mouthwatering buffet

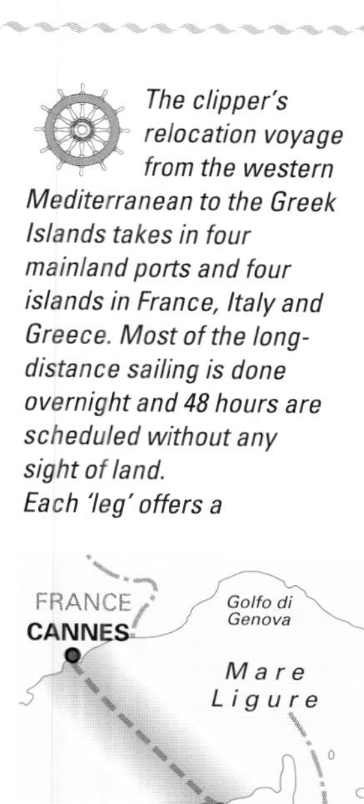

The clipper's relocation voyage from the western Mediterranean to the Greek Islands takes in four mainland ports and four islands in France, Italy and Greece. Most of the long-distance sailing is done overnight and 48 hours are scheduled without any sight of land.

Each 'leg' offers a comfortable mix of shipboard life and shore visits. Passengers board in Cannes and wake up approaching Calvi, where Nelson lost an eye and the battlements are still peppered with English cannonballs. The first optional excursion explores the Corsican hinterland.

The ship sails down the west coast, through the narrow Bocche di Bonifacio, to anchor off Porto Cervo, on the east coast of Sardinia (Sardegna). Although at the heart of a string of some 80 white-sand beaches, this modern village is noted more for its summertime classical concerts than its sand.

A dash across the Med to the Gulf of Naples (Golfo di Napoli) brings the ship off Sorrento, offering coach excursions to Mount Vesuvius (Vesuvio) and the ruins of Pompei.

Alternatively, hydrofoils and ferries cross to the Isle di Capri, with its luxury villas and phenomenal Blue Grotto.

The route to Sicily (Sicilia) deliberately passes close to the island of Stromboli around dawn so that passengers can have a grandstand view of any volcanic activity. The clipper then negotiates the swirling eddies of the Straits of Messina (Stretto di Messina) – a busy shipping lane prone to mist.

To reach Taormina – an attractive Sicilian town of flagstoned streets, sunny piazzas, frothing fountains, shaded cafés and wrought-iron balconies – the ship anchors off Naxos, a rather grubby seaside resort. Transport to town is scarce but the ship is patient.

It takes two full sailing days to reach the Greek island of Kýthira which appears as a barren moonscape, dotted with patches of scrub and dominated by an ultra-white hilltop monastery. But the inland town has real Greek appeal – white walls, blue doors, terracotta urns, vine-shaded patios, gossiping old men and donkeys motionless under the trees.

Nauplia (Náfplio), the old capital of the Peloponnese (Pelopónnisos), is an interesting town with a fort and back streets crammed with tavernas. Alternatively, a coach tour goes to the ruins of Mycenae and the 4th-century BC theatre at Epidaurus (Epidavros) – a scenic drive past many classical remains and wooded valleys of pine and olive.

Finally Piréas, bustling seaport and the first dockside mooring for nine days. The crew say farewell – and immediately start loading stores for the next voyage.

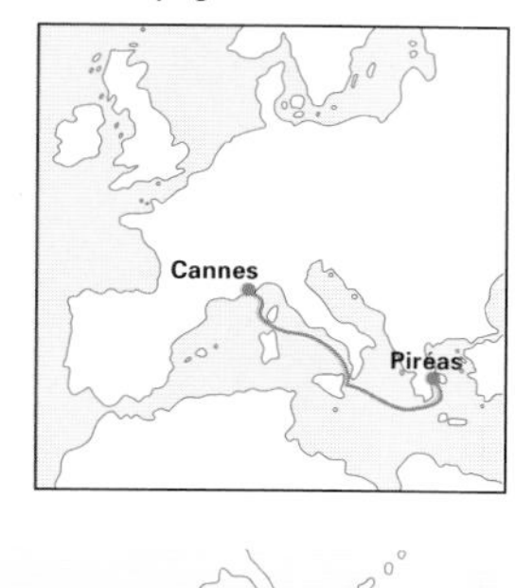

was laid out – smoked salmon, giant prawns, red caviar, *oeufs à la Russe*, glazed ham and a cornucopia of salads, with strawberries, fresh fruit salad and a variety of gateaux to follow. Clearly there was to be no scurvy or malnutrition aboard this sailing ship. The dining room itself was all mahogany, teak and gleaming brass, with Paisley patterned upholstery on the chairs and paintings of bygone clipper ships on the walls.

There was only one class. Passengers could sit anywhere at tables laid for four, six or eight and officers (including the captain) circulated. Formal dress was not required but we were asked not to wear T-shirts or shorts at dinner.

We decided to dine before unpacking, and afterwards went on deck, tots of rum in hand. The lights were gone but the stars were out and the moon continued to illuminate *Star Flyer's* white sails in an almost ethereal way. Eight bells sounded, the Watch changed and we went below to our comfortable cabin.

GREYHOUND OF THE SEAS

The old clipper ships used every inch of sail to clip hours off their journey times and so earn bonuses for early delivery of their cargoes. They were true 'greyhounds of the seas', pitching, rolling, tossing and tilting fearlessly in their unending quest for speed. The *Star Flyer* was

LEFT, *Captain Lickfett explaining the route on the chart table*

RIGHT, *Captain Lickfett checking the ship's bearing on the bridge*

more concerned about arriving at different ports of call on schedule and so used a 1,350hp engine, as well as sail, to control its pace and ensure that it did just that. On most days, anchor was dropped at around 9am, and weighed around 7pm.

On the first night the engine woke us, but thereafter we scarcely noticed it, not least because the cabins were well insulated. We had retired to find twin beds turned down, a card from Linfield, our Jamaican steward, laid beside a chocolate mint on the pillow and a sheet of advice about money tucked under the door.

Gdansk-born Captain Gerhard Lickfett, 40 years at sea with ten spent in command of sailing ships, was acutely aware that he had $40 million worth of responsibility under his command and left nothing to chance. As we prepared to drop anchor at our first port of call – Calvi, in Corsica (Corse) – he paced his bridge like a caged animal, megaphone in hand, eyes darting everywhere, watching the movements of the swarm of welcoming boats, checking instruments, taking compass sightings. 'Midships' he called to the wheelman, then 'stand by mainsail, stand by mizzen, stand by spanker.' Deck crew in blue-and-white-striped vests wound ropes around powered capstans to winch down most of the sails and adjust a cat's-cradle of rigging. Suddenly, to passengers watching from the sundeck, it was: 'Grab a rope, check a block and tackle, watch a sail – help the crew. Feel part of the ship.' A Klaxon hooted – and the tricky manoeuvring required to bring a 3,000-ton schooner to a stop in a small harbour was over. Officers (in shorts) prepared to lower away the tenders which would open up the delights of Calvi to the curious.

But first, there was 'Captain's Storytelling Time'. Captain Lickfett revealed himself as a rare species – a firm but fair disciplinarian as well as a humorist, raconteur and blackboard cartoonist.

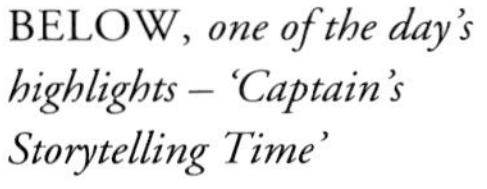

BELOW, *one of the day's highlights – 'Captain's Storytelling Time'*

With passengers grouped around the mainmast, he unfolded a personal version of Homer's *Odyssey*, mixed in with local sightseeing tips. It sounded a corny idea but proved highly amusing and became a highlight of each day.

PORTS OF CALL

Calvi harbour had a waterfront of attractive, canopied restaurants with twirling fans and a promenade where ladies still paraded their poodles. It was hot, so we took an air-conditioned excursion inland through the province where Columbus spent his infancy. The barren, hilly terrain was softened by great sweeps of *maquis*, vegetation made up from 2,000 different species of aromatic plant (Napoleon claimed that you could smell the *maquis* from offshore but his nose must have been better than ours).

One auberge, one olive-oil press, two hilltop churches and a Foreign Legion barracks later, we returned on the half-hourly tender to the ship. By now the sea had become a little choppy. The tender dipped, the *Flyer* rose and my right shin cut itself on the bobbing gangway. However, by now 'Happy Hour' had started and rum and hot 'nibbles' healed the soul, if not the flesh. I became acquainted with the Tropical Bar's additional attraction – a parrot called Dirty Harry (so named because of the state of his perch). The Tropical

ABOVE, Star Flyer *in full sail – classic elegance of a bygone age*

LEFT, *where better to enjoy the swell of the sea than the bowsprit net?*

ABOVE, *fancy yachts moor at Porto Cervo, a modern town on Sardinia's Costa Smeralda*

BELOW, *although there is still a faded charm to Sorrento, 20th-century tourism has taken its toll*

Bar, which opened at 10am, was the hub of shipboard social life and gave a grandstand view of the *Flyer's* afternoon watersports – snorkelling, aqualung training, wind-surfing, water-skiing and riding an inflatable yellow 'banana boat', towed at high speed.

At night the bar became the venue for dancing, an easy-listening and classical music quiz, a crew talent night and an unusual crab race. For this, a 16ft (5m) diameter chalk circle was drawn on the deck to mark the finishing line and five hermit crabs, fed and cared for since the ship was in Caribbean waters, were placed under a stainless-steel dish cover. Each crab had a number on its back and passengers placed $5 bets on which crab would cross the line first after the cover was lifted.

My wounded shin kept me aboard on Sardinia (Sardegna) at Porto Cervo, a modern town of terraces and boutiques developed in the mould of the Aga Khan, who owned the villa on top, but Helen went ashore. She related with pride, afterwards, how she had resisted a woollen coat with a fur collar at a sale price of 5,000,000 lire and a crocodile-skin handbag at 170,000 lire. Phew!

We sailed on to the Italian mainland. Sorrento, frankly, was a disappointment. From the sea it looked intriguing with its long line of Roman arches supporting the promenade and rococo houses with plenty of wrought iron, all 100ft (30m) above the beach, but whatever had attracted great poets and writers to the town in the past was now swamped with tourism. Two hours proved plenty. We probably should have taken the hydrofoil to Capri or the excursion to Vesuvio (Mount Vesuvius) and Pompei but we had visited them before and, by now, the ship had acquired 'home' appeal.

Dinner that night was for gourmets – asparagus, lobster tails and steak, meringues with cream and raspberries. As we passed Stromboli at 5.30am next morning, the volcano dutifully erupted on cue, belching flames 100ft (30m) into the air for the benefit of the 50 or so passengers on deck in their nightclothes.

On our way through the Straits of Messina, the captain's yarn urged us to listen for mermaids singing in the whirlpools. We did hear gurgles but, instead of mermaids, we saw a swordfish fishing boat with a look-out up its mast and a harpoonist straddling its enormous bowsprit.

Relaxing at Sea

The two full days at sea gave us a chance to explore the ship thoroughly, which proved to be a masterpiece of design. The four masts, each more than 220ft (67m) high, had a slight rake, adding to the schooner's elegance. The sundeck sported two swimming pools which were filled as soon as the ship dropped anchor. A watermaker, working on the principle of reverse osmosis, supplied 38 tons of fresh water a day. The bridge was set just below the sundeck so that we could lean over and watch the captain and helmsman at close hand during manoeuvres, while the bowsprit had a net slung underneath which allowed ten passengers at a time to watch the ocean rush by a mere 8ft (2m) below.

The Chart Room provided the greatest contrast between *Star Flyer* and the old tea clippers: radar, echo sounder, computer, auto-pilot, VHF radio, single red SOS button – even a Global Satellite Positioning System. This area apart, there were virtually no 'No entry' notices.

Our air-conditioned cabin had a porthole and mirrored walls to give an illusion of roominess and enough wardrobe and drawer space to please a fashion model. True, you needed to be a bit of a contortionist to control the shower but at least its spray stayed behind its curtain. Some cabins had baths. There was even a cellular-satellite telephone and a television.

Life on board was never boring. The crew were constantly fussing over the ship, touching up paintwork and re-varnishing, polishing brass or giving passengers demonstrations of sail-stowing, rope-handling, knot-tying – even vegetable sculpture and napkin-folding. At night, there were lessons in navigation by the stars.

When we reached the Greek island of Kýthira we found a cruise liner at anchor. It looked stunning and futuristic but its passengers were already being ferried back after only two hours ashore, whereas we had all day to get the 'feel' of being in a different country. The same was true at the ancient port of Nauplia, now Náfplio, where we similarly had plenty of time to explore.

By now a new atmosphere was pervading the clipper. It had become 'our' ship. We knew the lay-out in detail, we knew the crew by their first names and we understood the rigging. We had even grown to enjoy our rum out of plastic cups ('no glass on deck – captain's orders'), especially in the early evening when a low sun would turn everything golden. 'This is the time when the mermaids come out', Captain Lickfett persisted.

We knew we would miss 'Captain's Storytelling Time' and the cartoons on the blackboard. When we disembarked at Piréas, our final view was of him walking on tip-toe along an empty quay, arms outstretched sideways like a bird. Was he getting his 'land legs', we wondered? Or maybe rehearsing for his next story?

Practical Information

- The *Star Flyer* and its identical twin, the *Star Clipper*, alternate in different oceans – the Western Mediterranean, Eastern Mediterranean, Indian Ocean and Caribbean – and relocate at different times of year to find sunshine and favourable weather.
- Itineraries, dates and price details for all cruises and relocation voyages can be obtained from Fred Olsen Travel, White House Road, Ipswich, Suffolk IP1 5LL. Tel: 01473 292222. Fax: 01473 292345.
- Air-sea packages are available.
- There are six categories of cabin, depending on deck, position and degree of luxury. Cabins can be arranged to accept children. No pets are allowed.
- Dollars, dollar travellers' cheques or the usual credit cards are accepted but there are no currency exchange facilities on board. Excursions, drinks and insurance are the only 'extras'. The purser gives guidance on tipping at the final account stage.
- Light, cool clothing is recommended – walking shorts, bathing attire with cover-ups, skirts, light trousers and hats for ladies by day, sun dresses, slacks and blouses or comfortable knitwear for the evenings: at dinner, men should wear slacks and shirts with ties or polo necks.

LEFT, *view of Kapsáli's two bays from the castle in Kýthira town*

Corfu to Ithaca: a Greek Homecoming

ANTHONY SATTIN

ABOVE, *a red-figure Stamnos vase (c490* BC*) depicting Odysseus and his crew being lured onto the rocks by the Sirens*

Few stories from the past have as strong a hold on our imagination as the homecoming of the Greek warrior Odysseus, and few 'footsteps' offer such beautiful scenery as the journey from Corfu (Kérkyra) to Ithaca (Itháki). According to legend, after the Achaeans (Greeks) had destroyed Troy in revenge for the kidnapping of the beautiful Helen, it took Odysseus 20 years to get back home. Following Odysseus on the last stage of his journey led me across the Ionian Sea and, like the hero, my passage was far from direct – from Corfu to the mainland, to the island of Kefalloniá and then Ithaca. But as the modern Greek poet Constantine Cavafy wrote, 'As you set out for Ithaca hope your road is a long one, full of adventure, full of discovery.'

I was looking for a place 'with a fine harbour on each side, and only a narrow approach ...' and from where I sat, on the slope of a steep green hill above Palaiokastrítsa on Corfu's beautiful western coast, I thought I'd found it. Others have thought so too and have dug up the place in the hope of uncovering the ruins of King Alcinoös' palace. But nothing has been found. If this was the site of the palace, however, then this was where the great warrior's luck changed, for after many years of trials and tribulations, Odysseus had washed up in the country of a king who could help him.

When Odysseus insisted on returning to Ithaca, his wish was granted and his boat was filled with gold. The only gold I saw below me was in the colour of the sand and around the necks of some women lying on it, soaking up the sun. The only other link with the past was even more tenuous: a bar called Homer. But the hills were green, the monastery on the headland was old and venerable and the gentle Ionian sea sparkled like a gem set into the protected bay. It was a captivating view and one which, according to a sign in the café where I was sitting, had been savoured by kings of Greece 'as well as Kaiser, Tito and Nasser'. But no one stays here for long. Even after the king's beautiful daughter fell in love with him, Odysseus longed for his home on rocky Ithaca, for his wife and child. Like him, I dragged myself away. Odysseus' galley left from the harbour in front of the king's palace, but I had to cross the island to get on my boat. Homer tells us of no stops between Corfu and Ithaca, but Greek ferry schedules inflict Odyssean wanderings on to anyone wanting to travel to Ithaca, especially out of season: first to the mainland port of Pátra and from there, perhaps, if luck holds out, direct to Ithaca.

THE NIGHT BOAT TO PATRA

The *Fedra* was scheduled to leave Corfu port for Pátra at 9.30pm and I had been advised to be at the harbour in good time. The Greek military still occupy the solid block of the New Fort ('new' as in 'newer than the old one') built by the Venetians on a hill that dominates the port, and the US navy makes use of the facilities. Yet, for all this, the place felt abandoned and when I finally found an official, sitting in a back room over coffee and cigarettes, he knew no more than I did about where the boat would arrive.

At 10pm I was still sitting on my bag, watching the Albanian hills disappear into the night and indulging myself with the prospect of embarking on a sea journey, when there was a sudden flurry of activity. Officials hurried out of the customs

Homer doesn't give us a detailed itinerary of Odysseus' homeward journey, but he does give away information in his stories. Having lost all his men and escaped from seven years captivity on Gibraltar, Odysseus sailed alone across the Mediterranean until he was washed up on the island of the Phaeacians (Corfu), believed to be at Palaiokastrítsa. After staying there as a guest of King Alcinoös, he was provided with gifts and a swift boat home to Ithaca.

Corfu harbour is one of the most important in the area and, as a result, Corfu town is one of the most elegant of the Greek islands. The heart of the town was built in the 16th and 17th centuries, during the Venetian occupation, and still has a strong Italian flavour.

In the early 19th century the British took Corfu and developed the central Spianada, adding many of the grand civic or ceremonial buildings, including the fine neoclassical palace. The British influence continues here, with cricket played on the green and ginger beer served in the cafés.

Archipelagos, a marine and coastal management group backed by the WWF, has produced a leaflet laying out eco-friendly, sustainable walking trails on Ithaca and Kefalloniá. The leaflets, with maps and route descriptions, are available free of charge from hotels and travel agencies on these two islands.

Ithaca (Itháki) is unique in that it was made famous by a poem: Homer's Odyssey, not archaeological evidence, ties Odysseus to Ithaca and despite more than150 years of archaeological interest, little more is known about the whereabouts of Odysseus' home than Homer chose to tell us.

BELOW, *the natural harbour of Palaiokastrítsa, on Corfu, supposedly where Odysseus was washed ashore*

MAIN PICTURE, *rugged terrain and few beaches have saved Ithaca from mass tourism*

shed and cars drove along the quayside as the lights of the *Fedra* appeared out of the gloom. Within a few minutes she was upon us, towering over the dock. A few frantic minutes later, unloaded and reloaded, she was back out at sea, with me leaning against the deck-rail watching the brilliant lights of Corfu town and the giant illuminated crucifix above the Old Fort recede into the distance.

The *Fedra* had cabins, but my ticket was marked 'D/S class', which entitled me to either deck or Pullman chairs. For a while I considered wrapping up and sleeping under the black, star-encrusted sky, but it was still early in the summer and the night wind was chilly. Instead I settled down below in a large salon where the first of the summer's European backpackers and world-tour Americans were untying their bedrolls and hanging up their socks for the night.

At dawn, up on deck, a little stiff after a night in a chair, I found that we still had coastline to the east, only now it was Greek, not Albanian. And as the sun burned off the morning mist I saw land to the west. 'Is that an island?' I asked one of the crew, who was occupied in staring at the water breaking at the bows. 'Yes,' he replied without looking up. 'Itháki.' 'And are we stopping?' 'In Itháki?' He gave a little laugh. 'Why stop in Itháki?'

I remembered the passage in a poem by the late, great Greek writer C V Cavafy about not hurrying to Ithaca, about it being better if your journey lasts for many years, so that you are old and wise when you arrive – and so you don't ask or expect too much from Ithaca. The journey's the thing, he's suggesting, and I was inclined to agree as we passed Ithaca and, a little later, the bay of Mesolóngi where the English poet Byron had died.

At about 9am, in the beautiful channel which eventually joins up with the Corinth Canal, we suddenly turned towards land and found ourselves face to face with Pátra, its port dominated by a modern town, the town dominated by a medieval castle. On land I asked about the next boat leaving for Ithaca – 'Oh yes, there is one', I was told, 'early in the afternoon' – then headed into town for breakfast. In a yellow-stained cafe, a British couple I had seen on the *Fedra* were asking for croissants or toast. The waiter pretended not to understand them, while a neighbouring group of old men drank murky coffee and took hungry drags on their first cigarettes.

BELOW, *ferries jostle for position around the Ionian islands*

SAILING TO ITHACA

Sailing out of Pátra on the *Atomiko* at 1pm, I wondered what it was that attracted me to Ithaca. There was the romance of the story, of course, of the happy-homecoming, but most of all there was the adventure of travel, the chances of things happening, or not happening. I'd been assured there was a boat to Ithaca, but nothing is so easy in out-of-season Greek waters. Instead I was sailing to neighbouring Kefalloniá, close enough to my journey's end, I thought, for me to swim there if I had to.

As the sun was shining and the wind was down, most passengers happily forsook the bad coffee and worse television to stay up on deck. The *Atomiko* is smaller than the *Fedra* and has fewer facilities to offer, no cabins or well-stocked shop, but when the weather is good there's not much you need beyond a drink, a sandwich or some fruit, sun cream and sunglasses. Ahead of us the sea was as smooth as blue ice and the deep sky had been dissected by the trail of several planes. A young German couple put a Euro-rock tape in their portable music centre and, leaving my copy of Homer's *Odyssey* in my bag, I watched the land drift away and the islands move slowly towards us.

The sun was down when we pulled into the narrow channel that separates the green island of Kefalloniá from its smaller, more famous neighbour. Ithaca was now close enough for me to see its rugged glory, to see that Homer clearly knew what he was talking about when he called it rocky. Windy too, for the sea was suddenly choppy in the channel and I was glad when we pulled into the harbour at Sámi,

BELOW, *Kefalloniá, the largest of the Ionian Islands, and just a stone's throw from Ithaca*

OPPOSITE PAGE, *the Greek flag flies proudly from the deck of a local boat*

victim of a long line of calamities from attacks by Romans right up to an earthquake which levelled the place in 1953.

That night, at dinner on the waterfront, I was flicking through the *Odyssey* again when an American-Greek lady interrupted me. 'I couldn't help but notice what you're reading ...' 'I'm on my way to Ithaca,' I explained. 'And you?' 'Oh yeah. I'm going there. But you know what I've been told? There's no real evidence for any of that stuff. It's all just a made-up story.' (She made it sound like a television soap.) 'Ah, but what a story!'

Still, as I found out soon enough, she had a point: there is very little evidence beyond what Homer tells us and yet Homer was writing around 800 BC, some 400 years after the event. Facts are thin on the ground.

JOURNEY'S END

Without any noticeably long sandy beaches or gentle plains along the coast, Ithaca greeted me with rocky hills and a few stony coves. I found this rather strange. For when Homer made Odysseus into the model traveller, he turned Ithaca into the perfect end to a journey. Something like paradise, in other words. When we rounded the point of the green promontory, entered a fold in the hills and reached Váthi Bay, I understood why.

Encircled by hills, offering no view of the diamond-blue Ionian, Vathy, the island's main town, spreads itself around the far side of the bay. Its buildings are low-rise, classical in some of their façades. With relatively few young people, no topless girls, neon signs or disco bars, Váthi was unlike most other Greek island towns I had seen. In their place, foreigners from the flotilla of yachts tied up around the bay were drinking aperitifs on the terraces, before heading for the tavernas or walking around the small bay to eat fish by the water at Gregory's.

I was happy to be in Váthi and would gladly have stayed put for a few days, but there's no getting away from Odysseus on Ithaca. Later that night, after I had eaten and was having a drink on the front, staring at lights reflected on the still water, the American woman I had met the previous evening passed by. 'You're missing the spectacle. Look behind you. See the shape of the hills? Looks like a man sleeping, right? The locals

BELOW, *the stone bridge linking the two sides of the bay at Argostóli, Kefalloniá's capital, was built under British rule*

say that it looks like Odysseus and that one day he'll wake up.'

Even when Odysseus returned to Ithaca, his trials weren't over. Being a clever and cautious man – it was he, after all, who had thought up the Trojan Horse – Odysseus decided that first he should visit his old swineherd, a man called Eumaeus. Homer says the two men met at the Arethoúsa Springs. So before the shops were open or the yachts had cast off, I started the beautiful two-hour walk up a gentle slope, through olive groves, to the Springs.

On the way, I met men clearing the ground under their trees or, having finished, making their way home, and each of them stopped to ask where I was going. 'Ah, the *spring*,' a pensioner called Tassos said. 'From Homer... Odysseus... from the old times.'

The spring is as Homer describes it, just a water hole set in the rocks; but as Cavafy said, it's the journey to get there that is most pleasing. Hot from the walk, I did as Tassos had recommended, lowered the tin bucket and drank – as perhaps Odysseus had drunk – from the spring's cool, sulphurous water. Below me two German couples dived from their boat into the transparent turquoise of the sea.

With the help of his swineherd Eumaeus, his son Telemachos and the goddess Athena, Odysseus cleared the young suitors from his palace, was reunited with his wife Penelope and lived with her in their palace. Ever since the late 19th century, when the German archaeologist Schliemann found the remains of Troy by reading his Homer, archaeologists have tried to find Odysseus' palace. After excavating in the 1930s, a British team believed it lay north of the island, near the town of Stavrós, and Greek archaeologists continue to dig there. Meanwhile a professor from Washington University, St Louis, is digging at Aetos in the south of the island, near Váthi.

That night, having visited both sites, I again met the American woman. 'I don't know why I bothered', she complained. 'But both places fit in with Homer's description of the country around Odysseus' palace', I replied. 'But there isn't anything to see.'

The following day, starting out from Fríkes in the north of Ithaca, I caught the small ferry to the island of Lefkáda and, I sincerely hoped, a connection back to Corfu. I had the sun on my back, a breeze in my face, the smell of the sea in my nostrils and the American woman's words in my head. She would have liked to have seen Odysseus in his palace. But Cavafy had said 'don't hurry your journey at all' and that's how I felt about the archaeologists looking for Odysseus' remains. Take your time, there's no hurry to provide evidence because, for as long as Odysseus' palace is undiscovered, Ithaca remains an ideal, a place to set out for with more hope than expectation.

PRACTICAL INFORMATION

■ Summer and winter ferry schedules vary considerably. In summer, there are direct sailings from Corfu (Kérkyra) to Ithaca (Itháki). Out of season, there are daily boats on the main route from Corfu to Pátra (on the mainland) and often from Pátra to Kefalloniá and Ithaca.

■ To avoid peak-season crowds, and because of the ferry schedules, early or late summer is the best time to travel.

■ Overnight ferries between Corfu and Pátra have cabins, Pullman seats large enough to stretch out on and deck space. The most reliable ferry information is provided by local Port Authorities: Corfu (Tel: 661 32655), Ithaca (Tel: 674 32909), Kefalloniá (Tel: 671 22224) and Zakynthos (Tel: 695 22417).

■ There isn't much choice of accommodation on Ithaca, so the existence of these two well-run family hotels is paticularly welcome:
The Mentor, Váthi (Tel: 674 32433); Fax: 674 32293.
The Nostos, Fríkes. Tel: 674 31644/31100. Fax: 674 31716.

ABOVE, *the search for Odysseus's palace continues; a dig at Aetos, a couple of miles from Váthi*

LEFT, *ferries congregate at Váthi Bay, but most visitors are on their way again after an hour or two*

OPPOSITE, *colourful fishing boats moored at Fiskardo on Kefalloniá*

Secret Waterways: St Petersburg to Moscow

CHRISTOPHER KNOWLES

ABOVE, *the golden dome of St Isaac's Cathedral in St Petersburg*

In recent years, since the beginning of the dismantling of the Soviet Union, it has become possible to explore areas of Russia that were formerly closed to foreigners. Now it is also possible to do it in some comfort, thanks to the new policy of encouraging foreign investment. This river and canal route links two of the world's great cities – St Petersburg, Peter the Great's 'window on Europe', and Moscow, which remains, for all the changes taking place there, a reflection of the extremes of the Russian character. The journey takes passengers through the heartlands of ancient Muscovy and the monastery towns of the Golden Ring.

BELOW, *the Winter Palace, part of the complex of buildings called The Hermitage, was built for the Empress Elizabeth*

Contrary to the expectations of just about every passenger, each of whom had paid a not inconsiderable sum to be aboard the MV *Kirov*, moored on the Neva at St Petersburg (Sankt-Peterburg), the boat was a paragon of waterborne comfort. The cabins, it is true, were on the small side – but they were air-conditioned and everything, including the plumbing, functioned efficiently. The food was not gourmet but it was good and there was plenty of it.

Why should such doubts prevail? Because we were in Russia, a land infamous for food shortages and inefficiency. Anyone who had arrived only a few hours before at the airport that serves this city of extravagant palaces and burnished domes would have surely felt that little had changed since the era of the Brezhnev brotherhood. A full hour for the luggage to make its way from the aeroplane to the carousel? Another to pass through customs?

Yet, once the sleek *Kirov* (built in Germany and operated by the Swiss) was spotted in the twilight, all these frustrations were as nothing. The River Neva, forming a wide strait, separates the city into distant horizons where fantastical palaces line the waterfront against flat, murky skies. There is the tapering Nordic spire of the Peter and Paul Fortress, while near by is the battleship *Aurora* whose guns announced the beginning of the Revolution in 1917. The marzipan Winter Palace overlooks the river, in the shadow of the gleaming cupola of the reconsecrated St Isaac's Cathedral.

ABOVE, *the Peter and Paul Fortress, St Petersburg*

The next day was to be devoted to intensive sightseeing before setting sail into the Russian heartland. It is often said that St Petersburg is not really 'Russia', but that is not quite true. It is a manifestation of the fanciful rococo side of the Russian character, that flight of fancy that writes poetry or creates the *Nutcracker* ballet.

Peter the Great, first of the modern Tsars, wanted to bridge the age-old gap between east and west by building a 'window on the world'. So, he travelled the world to find ideas, in disguises that failed to conceal his 7ft (2m) frame, and hired architects and engineers from Italy and Scotland to build his new city.

AN UNKNOWN VASTNESS

That evening, as soon as the last bus had emptied its contents of Swiss, German or British passengers

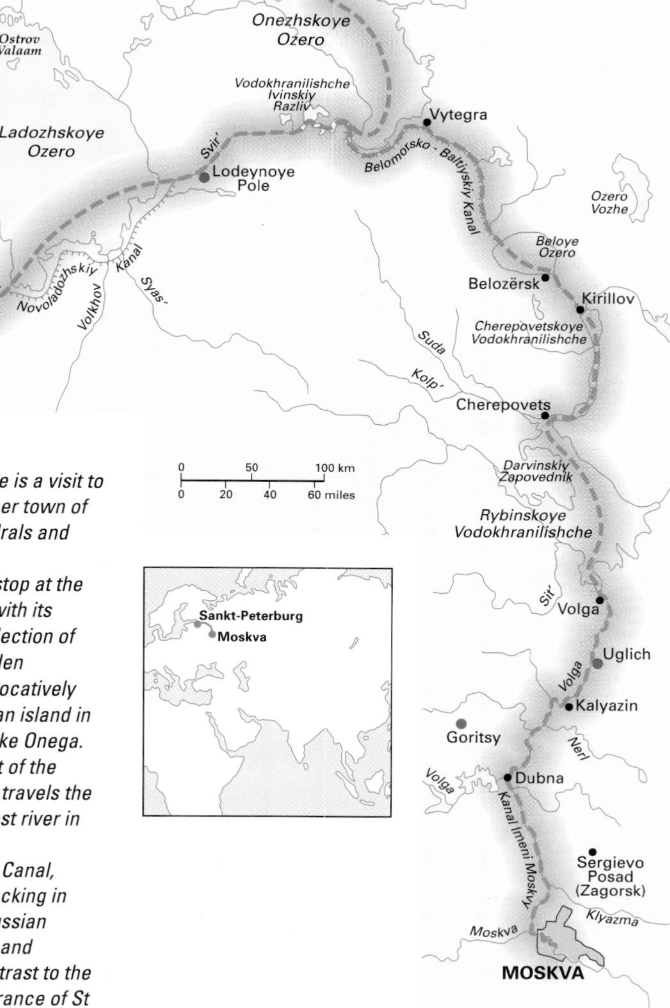

The journey begins in St Petersburg (Sankt-Peterburg), the capital of Russia from 1703 until the Revolution of 1917, after which its name was changed to Leningrad. It reverted to its original name in 1991.
There is much to see in the city itself – the Hermitage, containing one of the greatest art collections in the world, the Winter Palace, St Isaac's Cathedral, the palaces of Nevsky Prospect, the Peter and Paul Fortress – but there are also a host of magnificent palaces on the outskirts of the city, notably Petrodvorets, Pushkin and Pavlovsk.
The journey continues along the River Neva and on to the expanse of Lake Ladoga, the greatest lake in Europe. This is the beginning of the Russia of early history, a land of forest, small settlements – and the Golden Ring. This evocative name (not to be confused with the Golden Horde, refers to the Mongol era) has been conferred on the towns and monasteries that, with Moscow (Moskva), formed the heart of the state of Muscovy. Among those which lie on or close to the river are Goritsy and Uglich, where stops are made.
Sometimes there is a visit to Yaroslavl, another town of majestic cathedrals and monasteries.
There is also a stop at the island of Kizhi, with its magnificent collection of traditional wooden architecture, evocatively laid out across an island in the middle of Lake Onega.
For the final part of the voyage the ship travels the Volga, the longest river in Europe, and the Moscow–Volga Canal, before finally docking in Moscow, the Russian capital, austere and secretive in contrast to the colourful exuberance of St Petersburg.

ABOVE, *the Church of the Transfiguration is built entirely of wood – right down to the nails*

on to the quayside, the *Kirov* slipped its moorings and thrummed, in the gentle way of river craft, eastwards. We passed the baroque workshops of shipwright and dock, low buildings in ochre and white, and the nasty high-rise blocks of the Soviet era.

The next day we awoke to find ourselves in the middle of nowhere. That is to say, we appeared to have entered a seemingly boundless sea. This was, in fact, Lake Ladoga (Ladozhskoye Ozero), the largest lake in Europe. It is a measure of the immensity of this country that one is constantly discovering places, scarcely mentioned in the classroom, which turn out to be the largest of their kind, or at the very least the largest in Europe. Of course, we have all heard of the Volga, but what about the Yenisei and Ob, two of the longest rivers in the world? Who had ever heard of Lake Ladoga?

BELOW, *passenger boats moored on Lake Onega*

Above us in the early morning was an immense grey sky beneath which the calm lake waters took on a bronze sheen. Occasionally little thatches of green drifted by on the horizon, just some of the dozens of islands, many with ancient monasteries in varying degrees of degradation, that dot the lake. All day we cruised until finally a coastline loomed up and we entered the Svir River, which links Ladoga with Lake Onega (Onezhskoye Ozero).

The river, in common with much of Russia, is lined with tall straight trees. About 30 miles (48km) upstream from Ladoga we passed a substantial town, Lodeynoye Pole, which was again the object of Peter the Great's attentions. He found that the trees lining the river bank were ideal for ships' timbers. Here, from 1702, the first Russian naval vessels to appear on the Baltic Sea were built and although the Admiralty Shipyard in St Petersburg itself soon became pre-eminent, the shipyard here continued to build ships until 1830.

RUSSIAN ICONS

We pressed on. On board our own ship life continued at a leisurely pace. People stood at the railings for hours to watch the endless tree-clad vastness of Russia, which, if you allow it, can exert a hypnotic, restful fascination. In the early evenings there were talks about the history of Russia, followed by dinner. Then a general rush for the bar where the enigmatic Russian character was a subject which figured high on the agenda. In addition to the Europeans aboard there were a few Americans, and journalists from a variety of European and North American publications. Of course the main topic, concerning Russia, was whether or not democracy was likely to find a permanent place in Russian life. Another was the Gorbachev versus Yeltsin debate.

The next day, however, we were able to disembark. By now we had entered another lake, Onega, a mere 3,700 miles square (9,546sq km) and about nineteenth in size in the world. Our destination here was Kizhi, an island which became important in the 14th century as a stopping place on the trade route between the White Sea and the merchant capital of Novgorod to the south.

The populations of the 13 villages of Kizhi grew to the point where it became the administrative centre of an area that included a total of some 130 villages. Among the churches erected to cater to the spiritual needs of a devout population is the

Spasskaia, with tall towers that could be easily espied by approaching boats and then, most famously, the magnificent Church of the Transfiguration, built out of wood in 1714 and still resting on a knoll at the southern tip of the island.

It is the earliest of a group of three buildings on the hillock, the others being the Church of the Intercession built in 1764 and the bell tower of 1874. But the older church is the one that takes the eye, with its cluster of 22 silvery-grey cupolas rising to a height of 121ft (37m). Built on this windswept island, it is as evocative of the traditional Russian way of life as anything can be, even if it now forms part of a type of open-air museum known as 'ethnographic', a style much beloved of the old communist regimes that liked to pretend that the existence of the impoverished peasant, a slave to religion, was all in the past.

Rural Russians still live much as they have done for centuries. A few live on Kizhi whilst from time to time, from the boat, we were able to catch glimpses of their way of life in those few settlements that had sprung up among the trees. Elderly women dressed in black and wrapped in scarves – the archetypal 'babushkas'– still labour with buckets of water from well to cottage, against which piles of logs rest in readiness for the winter. The occasional beaten-up Lada sent up a cloud of dust from unmade roads, but more often it was a horse and cart, or a moped, riding the cratered tracks like a dinghy on a heavy sea. The dead stillness of winter, and the hermetic isolation, is easy to imagine.

ABOVE, *industry on Lake Onega: a huge log raft*

MONASTERIES ALONG THE CANAL

The ship now journeyed southwards, leaving the lake for the route that forms the Belomorsko–Baltiyskiy Kanal (White Sea–Baltic Canal), built in the 1930s by Stalin's convict labour, with appalling loss of life. We passed through dams that, small as they are by modern standards, were hailed as great achievements in their day, part of the 'electrification of the Soviet Union'. Locks, like stations, always attract crowds and there was a lot of banter between the passengers and the children who watched the opening and closing of the great dam door in wide-eyed wonderment.

Our next stops were to be at Goritsy to visit the Kirill-Belozërsk Monastery, and then to Uglich. The monastery lay at the edge of a small hick town of dusty roads and wooden houses. The people were friendly in a gruff way, with weather-beaten faces that could, with encouragement, break into dazzling smiles. The monastery itself, which dates back to 1397 and at the height of its influence owned 20,000 serfs, though complete, was in an advanced state of decay. The grass of the courtyards was long and unkempt, and the whole place wore an air of helpless desolation. But in its isolation close to the River Volga, it was touched with a wistful beauty that seemed wholly in keeping with the Russian nature.

Uglich, on the other hand, on the banks of the Volga, is one of the monastery towns that make up

The epitome of Russian architecture: the Cathedral of the Assumption and St Basil's Cathedral (inset), both in Moscow

the Golden Ring and therefore, unlike Goritsy, is used to a certain amount of tourism. The churches and the kremlin are in excellent condition – the Church of Dimitri's Blood, the Spasso-Preobrazhenskii Cathedral and the Uspenskaia Church, with its unusual tent-shaped roof, are all magnificent. But here the visitor is haunted by the black-market money changer and by the insistent souvenir seller. Not every seller is a pest, though – there are artists with their work on the pavement at their feet, unsure how to sell, and old women offering vegetables and home-made jam. A makeshift antique shop sells a motley collection of coins and religious artefacts of doubtful origin and authenticity.

Approaching Moscow

We walked back to the ship for the last leg of the journey, for that evening we would be in Moscow. Following the tree-lined Volga before branching south on to the Moscow–Volga Canal, the approach to the Russian capital was like entering a city by train, a gradual proliferation of man-made objects culminating in suburban settlements, and then the great city itself.

Moscow is far from St Petersburg in every sense. To be sentimental, one might say that one represents the Russian heart, the other – Moscow – the Russian soul. Moscow is a city of intrigue and secrets. Like in imperial China, power lies behind a wall, in the red-brick Kremlin, to and from which there is a regular coming and going of black limousines, orchestrated on Red Square by whistling, baton-waving policemen. The city itself has lost its walls, of course, but with its ring of towers, Stalin's skyscrapers that are Gothic parodies of the Empire State Building in New York, it manages to convey the compact insularity of a medieval town.

The inevitable addition of 20th-century frippery cannot detract from Moscow's fascination, which is the fascination of history, power and not a little beauty. A walk around any of the central streets away from the main sites is a journey to the Russia of the Tsars. To have visited an Orthodox church at dusk, filled with vacillating candles and the profound tones of the bearded priest, is as moving an experience as it is possible to have.

So, we saw what we could of Moscow; and then we went to visit Zagorsk (now rechristened Sergievo Posad), another monastery on the Golden Ring, with its star-spangled cerulean domes standing proud of the walls which encircle it like a fortress. Here, the burial place of Boris Godunov and one of the holiest shrines of Holy Russia, the churches have been opened again to worship, and the pilgrims, denied for so long, give thanks in their hundreds.

It was impossible not to look back lingeringly upon the bulbous skyline of domes and crosses as we made our way back to Moscow and the river for one final meal on board the *Kirov*, and one final furious debate in the bar about the future of Russia. Then to bed, to prepare to do battle next day with the frustrations of air travel; and to wonder at the scale of Mother Russia, and to wonder how she will cope with no Tsar and no Party.

Practical Information

- Modern ships such as the MV *Kirov* are often owned by a consortium of Russian and overseas companies which means that a high standard of comfort and food is combined with local knowledge.
- The service tends to operate throughout the months of late spring, summer and early autumn (ie from May to October). Russia's waterways are frequently unnavigable the rest of the year because of harsh winters. During these months there would typically be at least three departures each month, and also services in the reverse direction.
- The actual journey takes six days, although there are additional days spent in St Petersburg and Moscow.
- Whilst all cabins in ships of this standard face outside, have two bunks, air-conditioning and shower rooms, there are different prices according to whether your cabin is on the upper, main or lower deck.
- Facilities on the ship include bars, lounges, a sauna, a clinic, a shop and a restaurant.
- The *Kirov*, which carries 280 passengers, was built in Germany in 1989 and is operated by ICH, a Swiss cruise and hotel management company. Tickets for this trip can be obtained from Voyages Jules Verne in Britain (21 Dorset Square, London NW1 6QG. Tel: 0171 616 1000) or Berrier Enterprises in the USA (1 Sutter Street, Suite 308, San Francisco, CA. Tel: 415/398 7947).

LEFT, *the Kremlin, Moscow's magnificent fortress and seat of government*

Discover the Real Alaska

SHIRLEY LINDE

ABOVE, *traditionally dressed native Tlingit boys from Sitka*

The SS *Universe Explorer* cruises the Inland Passageway of Alaska, leaving from Vancouver, British Columbia, and travelling north to Seward and back again, winding its way through a spectacular land of eagles, whales and glaciers. The emphasis is on learning about the environment and the culture, but there are also plenty of opportunities for sightseeing and embarking on exciting shore excursions.

BELOW, *downtown Vancouver borders Burrard Inlet, the city's busy international harbour*

If you want to explore the real culture of Alaska, consider the trip offered by World Explorer Cruises. Instead of mink stoles, tuxedos, gambling and glitz, passengers dress in parkas and wool sweaters, marvel at spectacular glaciers, visit old gold-mining towns and Indian villages, and even attend on-board lectures about Alaskan life and culture.

The ship leaves on Tuesday afternoons from Vancouver but I arrived the night before to make

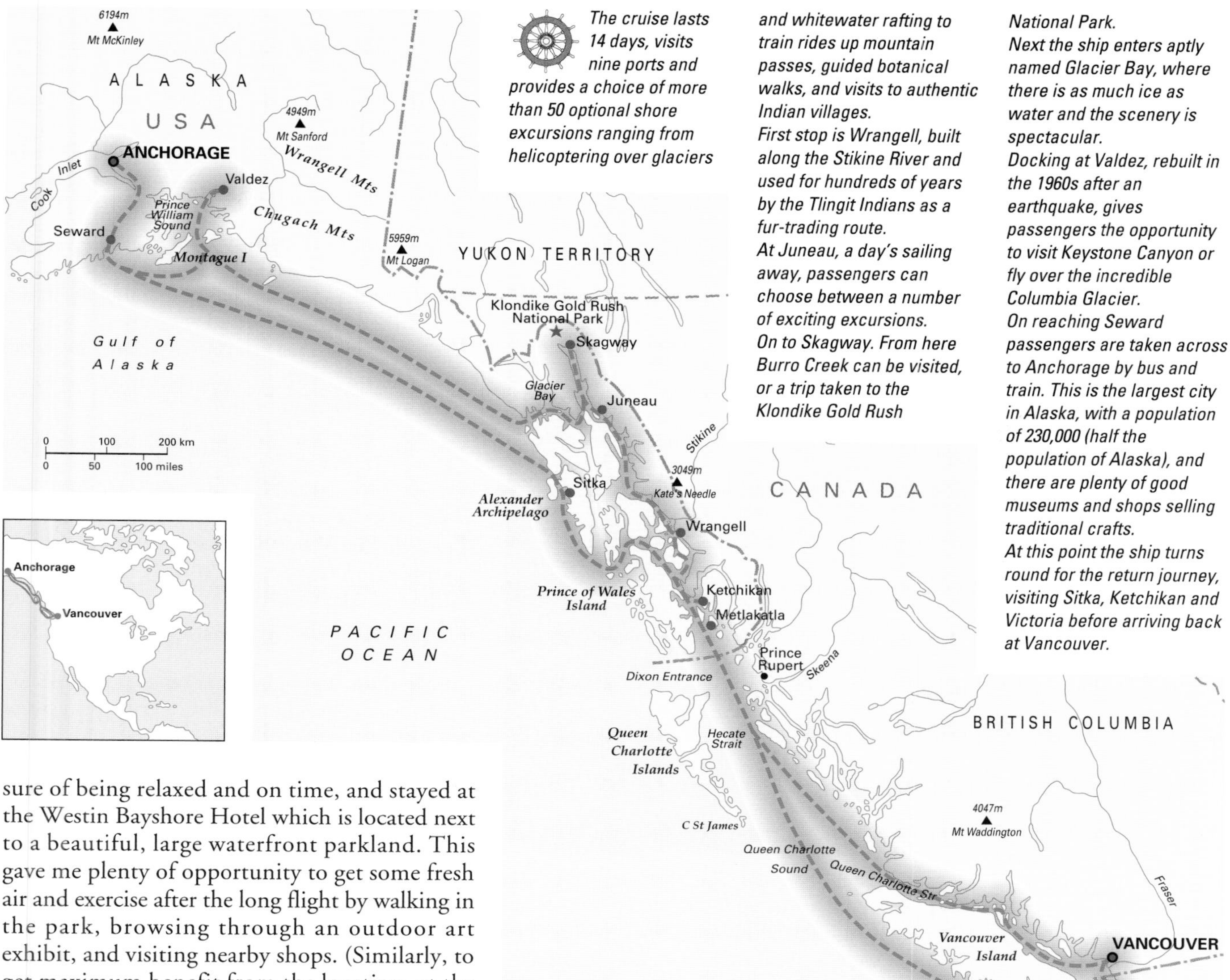

The cruise lasts 14 days, visits nine ports and provides a choice of more than 50 optional shore excursions ranging from helicoptering over glaciers and whitewater rafting to train rides up mountain passes, guided botanical walks, and visits to authentic Indian villages.
First stop is Wrangell, built along the Stikine River and used for hundreds of years by the Tlingit Indians as a fur-trading route.
At Juneau, a day's sailing away, passengers can choose between a number of exciting excursions.
On to Skagway. From here Burro Creek can be visited, or a trip taken to the Klondike Gold Rush National Park.
Next the ship enters aptly named Glacier Bay, where there is as much ice as water and the scenery is spectacular.
Docking at Valdez, rebuilt in the 1960s after an earthquake, gives passengers the opportunity to visit Keystone Canyon or fly over the incredible Columbia Glacier.
On reaching Seward passengers are taken across to Anchorage by bus and train. This is the largest city in Alaska, with a population of 230,000 (half the population of Alaska), and there are plenty of good museums and shops selling traditional crafts.
At this point the ship turns round for the return journey, visiting Sitka, Ketchikan and Victoria before arriving back at Vancouver.

sure of being relaxed and on time, and stayed at the Westin Bayshore Hotel which is located next to a beautiful, large waterfront parkland. This gave me plenty of opportunity to get some fresh air and exercise after the long flight by walking in the park, browsing through an outdoor art exhibit, and visiting nearby shops. (Similarly, to get maximum benefit from the location, at the end of the cruise I spent several days at the Friday Harbour Inn in the San Juan Islands and took a cruise up the Columbia and Snake rivers with Alaska Sightseeing Cruise West, but that's another story.)

ENJOYING SHIPBOARD LIFE

In the late afternoon, with the ship's band playing, the SS *Universe* (this was the ship I travelled on, now replaced by the SS *Universe Explorer*) sailed under Lion's Gate Bridge westward out of the bay into the sunset, a rosy preview of the good days ahead.

The first evening was spent getting to know the layout of the ship, becoming acquainted with a number of fellow passengers, finding the dining room to partake of our first fresh seafood of the trip, taking the first walk around the promenade deck (still viewing the lingering sunset), then discovering the jazz pianist in the aft lounge. This was followed by dancing in the commodore lounge and talking to yet more new acquaintances, while reflecting on how making new friends is one of the bonus features of any cruise.

The next day was spent at sea and passengers were able to take their pick of aerobic classes, a film on Alaskan scenery, or the first lectures on Alaskan history and anthropology (later there were lectures on bird and animal life, marine biology, geology and art). You could really delve into the culture, history and background of the people, the wildlife and the environment by visiting the library with its 15,000-odd books, and there were also files of information on where to shop, eat, visit and walk in each port.

During the cruise there was a get-acquainted party for passengers travelling alone and a welcome aboard cocktail party hosted by the captain (one of only two dressy evenings). Entertainment included a jazz pianist, a string quartet, a flautist and a classical guitarist, a dance band, plus bingo,

ABOVE, *Pacific walruses dead to the world. Lectures on Alaskan wildlife are a highlight of this cruise*

ABOVE, *the port of Juneau on the Gastineau Channel*

BELOW, *where visitors go to find out about Skagway and the surrounding area*

bridge tournaments, ship horse-racing, a costume night, a talent night, films and dance instruction. Children were well catered for, with a computer centre, computer classes and shore excursions of their own.

In between, you could partake of as many meals as your body would allow: early breakfast on the aft deck, regular breakfast, late breakfast, lunch on deck or in the dining room, dinner, plus mid-afternoon and late evening snacks in case you were still hungry; and then there was the midnight buffet. Alternatively, you could get a massage, have your hair cut, or put in a piece or two on the huge everybody-pitch-in jigsaw puzzle in one of the lounges.

All shipboard facilities, including the dining room, promenade deck and commodore lounge, had designated smoking and non-smoking areas.

World Explorer Cruise Lines, by the way, is one of the cruise lines that provides dance hosts on board to dance with women travelling alone, or whose husbands don't dance.

SIGHTSEEING OR ADVENTURE

The first port of call on our outward trip was Wrangell, one of the oldest towns in Alaska, with a population of about 2,000. It is one of the mildest of Alaskan towns, with temperatures seldom going below freezing even in winter, but can be very wet.

One of the highlights of Wrangell is the petroglyphs on the beach, primitive carvings in stone estimated to be more than 8,000 years old. If you take along some large sheets of paper you will be able to make rubbings using fern growing near the beach. Locally gathered garnets can be bought from children at the dock for next to nothing.

The next day's port of call was Juneau, where passengers could choose between two hours rafting on a river at the foot of a glacier, visiting an old gold mine and panning for gold, flying over the Juneau Ice Fields – 1,500 square miles (3,870sq km) of solid ice – or visiting a lodge in the wilderness and being served a fresh salmon dinner baked over an alderwood fire with wine chilled in glacier ice. Alternatively, you could walk around town and visit the Alaska State Museum, the antique-filled Alaskan Hotel or the original gold-diggers' hangout, the Red Dog Saloon.

The next day we reached Skagway, which has a population of just 768. Here the options were to visit Burro Creek by boat to see a private salmon hatchery built by a man trying to help preserve Alaska's depleting salmon, or to take a ride over the mountains and through the valleys of the historic Klondike Gold Rush National Park, imagining the thousands of men who walked these rugged paths in search of gold. In fact, many died and few got rich.

From then on we began to see wildlife: back on board a whale was sighted off the starboard side and an eagle could be spotted soaring overhead. All along the inland waterway were deep uninhabited forests and snow-capped mountains, their peaks towering on both sides of the ship, dwarfing

it. When we talked to one of the lecturers at breakfast on the aft deck, while it was still misty outside, he spoke about the problem of people being attacked by bears as they wandered along some of the trails. 'Still?' 'Definitely.' This is the wilderness you have come to see, but you make a note to stay out of isolated areas.

LEFT, *hopeful prospectors sifting for gold in Klondike just after the turn of the century*

GLACIER COUNTRY

Early next morning the ship slowed to a crawl – we were in the ice-filled waters of Glacier Bay. We rushed on deck to see the mist rising from the slushy, undulating water, icebergs and ice chunks afloat as far as the eye could see. Some had seals on them, lounging lazily, while seagulls and puffins flew overhead.

The ship inched forward to within ½ mile (1km) of the glacier and we were glad to be aboard one of the few ships that are allowed this close to the glacier. Bundled in a parka against the cold and drizzle, I stared in awe at the wall of ice hundreds of feet high, jagged, crevassed. This ice was laid down thousands of years ago – some of it by snow falling at the time of Christ! As we watched, huge sheets of the icy wall exploded away with a thunderous roar and a rifle-like crack that echoed through the vastness, and then ice fell slow-motion into the sea causing great splashes and rocking waves – a truly awesome sight.

After spending the next day at sea, we docked in Valdez, a town destroyed by an earthquake in

BELOW, *the inhospitable but beautiful waters of Glacier Bay*

PRACTICAL INFORMATION

■ The SS *Universe* has been replaced by the SS *Universe Explorer*, the new ship of World Explorer Cruise Lines. She is 617ft (188m) long, 84ft (25m) wide, with a tonnage of 23,500 and an 18-knot cruising speed. Passenger capacity is 739.

■ The SS *Universe Explorer* sails for World Explorer Cruises in Alaska during the summer, but for most of the rest of the year is operated by the University of Pittsburgh and the Institute of Shipboard Education as a floating university, called Semester at Sea, on round-the-world voyages.

■ It is air-conditioned and has stabilisers and an elevator service. There are eight decks, several lounges, a ballroom, a piano bar, a dining room with two sittings, a gift shop, a hair salon, medical facilities, a free laundry and a library stocked with 15,000 volumes.

■ All cabins have private bath facilities; many staterooms have berths for a third or fourth person.

■ A valid passport is required, plus a driving licence or a birth certificate for proof of citizenship.
For further information call World Explorer Cruises. Tel: 415 393 1565, or the free number 1-800-854-3835. Also Cruise World, Sydney, NSW. Tel: 2 9966 1677. Alternatively, contact your local travel agent.

ABOVE, *Anchorage, at the head of the Cook Inlet, backed by the Chugach Mountains*

1964, but – Alaskans are strong – simply rebuilt again 4 miles (6km) further inland. From here you could take a tour up Keystone Canyon to see waterfalls cascading from the canyon walls – world-class climbing is done here – and see the terminal of the Alaskan pipeline, where oil is piped across 800 miles (1,280km) of wilderness to tankers waiting in the Valdez harbour. Most passengers also took half-hour flights by helicopter or float plane over the Columbia Glacier. The view was breathtaking as we hovered over cliffs of ice and looked down on pools coloured an incredible penetrating turquoise blue and jagged, yawning, mile-deep crevasses. There was ice as far as the eye could see, centuries old, still grinding mountains and carving valleys as it did in the Ice Ages. It is sobering to reflect that the glaciers have been periodically advancing and retreating in this area for some 12 million years!

BACK DOWN TO CANADA

A day later the ship reached Seward, the most northerly point of the cruise. From here we boarded a bus which took us across to Anchorage.

There are excellent museums and art galleries here and a chance to do last-minute shopping for *mukluks* – to keep your feet warm – local jade or scrimshaw ivory. You can buy authentic prehistoric artefacts in many places, but we found some particular treasures in a shop called Boreal Traditions in the Hotel Captain Cook.

Two days later, now heading back down the coast again, we reached the town of Sitka. This is the ancestral home of the Tlingit tribes, scene of their last stand against the Russians who settled on Kodiac I in 1784. The United States bought Alaska from Russia for $7.2 million in 1867. We saw many totems and much magnificent basketwork in Sitka's Sheldon Jackson Museum before taking a peaceful walk along a trail through towering forests.

The last stop in Alaska is Ketchikan, where you could see 300 original hand-carved totems in the Totem Heritage Center, or help paddle a 37ft (11m) canoe across a wilderness lake, arrange for a private fishing trip, or take a float plane to Metlakatla, the only Indian reservation in Alaska. Traditional native arts and crafts are available for viewing and purchasing.

Scheduled for the next day was a stop in Canada's charming city of Victoria before the ship returned to Vancouver. We left the ship with a new understanding of the word wilderness and of the culture, tradition and people of the biggest state in the US, isolated from the rest of the nation by Canada.

■ Hooded rain gear to wear over a sweater or parka is essential.

■ The SS *Universe Explorer* now also makes two Caribbean cruises in December and January. The same emphasis on culture and learning is part of the Caribbean voyage as it is on the Alaskan voyage, with in-depth lectures and guidance from experts, both aboard and ashore.

■ The 15-day, 14-night cruise embarks from Nassau to Montego Bay, Jamaica; Cartagena, Colombia; Cristobal, Panama; Limon, Costa Rica; Puerto Cortes, Honduras; Playa del Carmen, Mexico; and returns to Nassau.

BELOW LEFT, *the wilderness of the forested tundra and the Alaska Range is only a few miles north of Anchorage*

BELOW, *one of the many totem poles from the north coast that can be seen in Vancouver's Stanley Park*

ABOVE, *huge and colourful, traditional paddle-boats ply the Missouri/Mississippi*

Down the Mississippi to New Orleans

ROB STUART

If ambition is a species of madness, then sailing 2,600 miles (4,160km) down the longest river system in America in an 11.5-ft (3.5m) dinghy powered by a 4-hp engine, *and* with both rivers in flood, was a case either of sheer insanity or, as a fisherman in Pierre, South Dakota, colourfully put it, having 'balls of steel'! Known as 'Big Muddy', the Missouri is the longest river in the United States, beating the Mississippi by 118 miles (188km). It is also one of the most dangerous, with 'more grief to the mile than any other waterway in America', as Mark Twain put it.

Crossing Lakes Sharpe and Francis Case, on the first leg of the journey from Pierre, was a salutary introduction to the vastness and solitude of the Mid West. In the words of the writer Charles Olson, it is a 'geography at bottom, a hell of wide land from the beginning'. Apart from the treacherously half-submerged cotton-wood tree stumps, the result of a series of dams built on the upper reaches of the Missouri to regulate flooding and irrigation, all we saw during our four-day crossing of these lakes were interminable rolling prairies, bald-headed eagles and the occasional deer. But the rigours of this part of the journey, harsh weather and thunderstorms

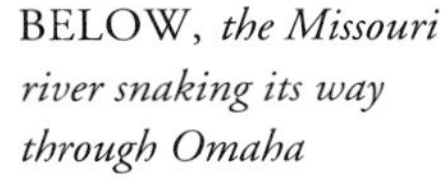

BELOW, *the Missouri river snaking its way through Omaha*

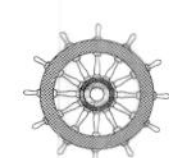

About 180 miles (288km) long, Lakes Sharpe and Francis Case are situated in a vast landscape of prairie wilderness inhabited by bald-headed eagles, coyotes, deer and wild turkey.
Visiting Pierre and Chamberlain is like taking a step back into history, and the Indian Reservations should not be missed.
From Fort Randall Dam down to Yankton there are glimpses of the old natural river as the early pioneers must have seen it – pioneers such as Lewis and Clarke, who first mapped the Missouri back in 1804.
The Missouri's pioneering history belongs to towns unbowed by economic recession, such as St Joseph, Waverly, Glasgow, Jefferson City, Washington and St Charles – and their hospitality is unsurpassed.
Of all the major cities along the Missouri and Mississippi, St Louis is the most attractive – a marriage between a feisty southern belle and a mid-western homesteader.
Marvel at the towboats, some of which push upward of 40 barges, and are the best testimony to the awesome size of the Mississippi south of Cairo, where the Ohio River joins the great river. Marvel at a respectful distance, however.
After the remote, backwoods towns of Hickman, New Madrid (old Madrid was buried by an earthquake), Tiptonville and Caruthersville, Memphis comes as a cultural shock. Vibrant, brash and proud of itself, it lays claim not just to being the birthplace of Rock and Roll, but also the Pork Barbeque capital of the world.
Helena, Greenville, Providence and Rosedale have a homespun but wretched look about them, yet in many ways typify this part of small-town rural America.
Vicksburg, perched on a steep bluff, is one of the most interesting towns on the lower Mississippi, not least because of its Civil War history and memorabilia. Outside the Old Court House Museum, the Confederate flag still flies.
'What goes in America, goes on in New Orleans.' Shamelessly exuberant and dynamic, its cultural diversity is nowhere better seen than in the French Quarter. But rampant commercialism is the Quarter's downside.

included, were offset on more than one occasion by the help we received: at Pierre itself; from a family of Sioux Indians at Lower Brule Reservation; and from the people of Chamberlain, a small but prosperous frontier town on Lake Francis Case.

My companion Tony Graham and I decided that the trip was to be an adventure, that we would go Huck Finn-style – travel light, sleep rough and live on the rivers. The heaviest items of equipment we took were two 5-gallon (19 litre) petrol tanks, a Coleman stove, a two-man tent, and the excellent US Corps of Engineers maps. These maps were indispensable, with their detailed and accurate illustrations of the rivers, lists of campsites and mileage charts.

After a portage at Fort Randall Dam, we made rapid progress on the swollen river down to

LEFT, *the Kansas City skyline is typical of many American cities*

Niobrara, and the small riverside community of Lazy River Acres. Harassed by storms, wind and rain, we pulled up here and stayed for three days as guests of Willard and Lucille Kuchar. Niobrara, once situated on the river, was so frequently flooded the Corps of Engineers finally put the whole town on low-loaders and took it to its new site on higher ground.

HAZARDS AND HOSPITALITY

When the weather eventually improved, we set off again – accompanied by huge amounts of debris in the river such as uprooted trees, parts of clapboard cabins, the odd truck – for Yankton where the official navigation route begins. There we expected to meet towboat traffic, but none appeared. At Omaha, where we stopped for supplies, we were told, to our great relief, that the towboats had been temporarily suspended because of the flooding. Of all the hazards on both rivers, the towboats were the ones we feared most .

Omaha and Kansas City are the two largest cities on the Missouri, but both were disappointing. Their block-style generic architecture on gridline design, so typical of American cities, is austere and featureless. 'Where's the life in this place?' we asked a passer-by in downtown Omaha. 'Oh', she replied, somewhat sadly, 'that moved out of town long ago.' More interesting are the small towns – Nebraska City, St Joseph, Waverly, Glasgow, Chamois, Washington. Though blighted by economic recession, these towns helped pioneer early American history, thriving on the bustling steamboat trade, commercial shipping and the westward-bound migrants. Most attractive of all is St Joseph with its ante-bellum and Victorian architecture of gracious merchant houses and civic buildings.

RIGHT, *St Louis's Gateway Arch dominates the city and riverfront*

BELOW, *the Capitol building in Jefferson City was constructed between 1911 and 1918 from Carthage marble*

Contrasting with the hills and high bluffs of Nebraska and Iowa, south of Kansas the river broadens out into spectacular flat arable land – though, on this occasion, apart from seeing marooned and sandbagged farms, little else was visible because of flooding. Resourcefulness and a little daring, however, got us through this stretch of the river. We camped on islands, slept on porches and in gardens of riverside cabins, or, in the case of Glasgow, literally sailed the floodwater through the campsite into town. Here we took refuge at Larry's motel which, like most motels in this part of America, was cheap, serviceable and clean. At Hartung's restaurant, owned by Gloria (for whom we shall always have special affection), we ate and drank well. America is renowned for its hospitality, and the people of Glasgow showed us just how generous it can be.

Restored, with freshly laundered clothes and full stomachs, we pressed on towards Jefferson City, capital of Missouri with its state building, featuring celebrated murals, that looks remarkably like a miniature White House, and Washington. Between Glasgow and Jefferson City it was 86 miles (137km) but with the river running at 12mph (21kph) we easily made it in a day. There we were warned about copperhead and moccasin snakes, the latter being the most venomous, that lurk in the driftwood on the banks. Thereafter, wherever we camped we always lit a large bonfire to ward off not just snakes, but also the increasing numbers of mosquitoes.

Two days off reaching the Mississippi River, we stayed at Washington to refuel and service the outboard engine. This old frontier town was settled originally by German immigrants (after the failed 1848 German Revolution) and I noted in the B&B in which we stayed an old German motto: 'One's own hearth is precious' – a sentiment reflected in the extraordinary civic pride of this small and bustling town. From the window of our large and elegant Victorian room overlooking the Missouri, I could appreciate what a welcome sight Glasgow must have made to those unhappy immigrants fleeing Europe's tyrannies.

WHERE THE RIVERS MEET

At 7.30, on a bright June morning, we celebrated our arrival at the confluence of the Missouri and Mississippi rivers with what was left of our Wild Turkey whiskey. After a short journey along the canal we moored up in St Louis beneath the famous Gateway Arch. Our celebrations were short-lived, however, when three US coastguards boarded our boat and placed us under 'hospitable arrest'. As one coastguard sternly but politely

Adam's Mark
HOTEL
ADMIRAL

explained: 'All pleasure boats are banned indefinitely from the Mississippi because of the flooding.' It was news to us.

Our enforced stay at St Louis gave us an opportunity to explore the place, perhaps the most attractive of all the cities along the two rivers. It is a mixture of southern hedonism and mid-western reserve, the boisterous street jazz at Leclade's Landing contrasted surprisingly with the subdued and conservative atmosphere of Central West End, a Parisianesque quarter of restaurants and shops. But the place we kept returning to was Union Station, recently converted to an arcade with street entertainment, good bars and restaurants, shops, and interesting people.

St Louis is named the 'City of Smiles' (so I was told), and after two days we too had reason to smile as the coastguards arranged for us to travel to Cairo – a two-day run – on the *W J Barta*, a towboat headed down to New Orleans.

From the bridge of the towboat I was given a view of the Mississippi that seemed to broaden out and engulf half of America. I was also advised by the captain of the perils ahead of us: 'With the river in flood the water round the wing-dykes will be fierce – lot of boils and whirlpools. And take care to give oncoming tows a wide berth. Their bow-waves can reach 5ft (1.5m) or more.' We saw what he described when we eventually set sail again from Cairo. In fact, such was the pressure of the water against the dykes it made the river look as if it was flowing upstream!

The increasing heat and humidity, together with long, often tedious, days on the river, brought acute exhaustion. We sweated profusely, ate little and dozed a lot. What's more, around

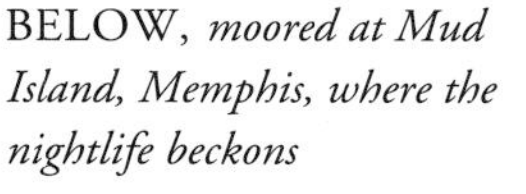

BELOW, *moored at Mud Island, Memphis, where the nightlife beckons*

Hickman, where we camped on scrubland below the town's flood wall, I contracted a nasty rash of poison ivy. Eventually checked by wych-hazel and calamine, it was marginally less irritating than the colonies of mosquitoes that continually besieged us, and for which there was *no* repellent.

Outside Wilson, an old plantation town where we camped one night, I noted in my diary: 'There's no such thing as a quiet night on the Mississippi. We can barely sleep for the noise of bullfrogs, mosquitoes, crickets, quails – all humming frantically and deliriously.'

DOWNSTREAM FROM MEMPHIS

The Yacht Club at Mud Island, Memphis, provided an excellent (and free) mooring: the manager was sympathetic to 'lone travellers'.

That night, Beale Street was our destination. At the Blues City Café we listened to Mama Toy – part Indian, part Irish, part Dutch, and in her 60s. She sang 'Catfish Blues' karaoke-style. Best of all was Silky O'Sullivan's where we listened to two virtuoso honky-tonk pianists. But we were too late for the black musicians we really wanted to hear – W C Handy, Booker White, Muddy Waters. Instead, we were treated to 'Little Boys' Blues' by five white lawyers; and Osaka Hurricane, a Japanese guitarist. Beale Street, like Bourbon Street in New Orleans, plays more to the uneasy tune of tourism now than for the aficionados of traditional jazz and blues.

ABOVE, *for music-lovers, no visit to Memphis is complete without a trip to Beale Street*

The recent proliferation of gambling boats along both rivers, notably Harrah's and Lady Luck's, offers cheap food and accommodation and, on some, free drinks if you're playing at the tables or on the machines. Tony, I discovered, was an adept blackjack player, and on more than one occasion his winnings helped top up our dwindling funds. But not all the casinos are welcome in the towns they're moored at. At the Spinning Wheel Bar at Greenville, for instance, the owner was bitter about them: 'If you can drink free beer at the casinos, what hope do we have?' To river travellers like us, however, they are a boon.

And for a while at least our luck held out. At Vicksburg, perhaps one of the most historical towns of the southern states and certainly one of the most distinctive, situated as it is on a high bluff overlooking the confluence of the Mississippi and Yazoo rivers, we met Tommy and Dora who offered us a lift on their cruiser down to Natchez. Tommy described himself as a 'River Rat', and had been travelling the Mississippi and its tributaries for more years than he cared to remember. Tommy and Dora were headed for Red River. What was the Mississippi's fascination for him? 'Out here is just damn big territory – how nature intended America to be!'

Around Baton Rouge we entered the industrial part of the river, festooned with gantries, factories, pipelines and large docks for sea-going freighters. The air was sulphurous and dark – and not just because of pollution. A tropical storm was brewing. When it broke it hit us with such ferocity that rain ricocheted off the river into our boat, barrels

BELOW, *Shirley House is the only surviving structure from the Civil War in Vicksburg National Military Park*

of thunder and lightning rolled deafeningly from the sky and the river turned a ghostly effervescent white. Fearing a collision with a towboat, we took shelter under the canopy of a sunken launch in a mosquito infested bayou. While bailing out our boat, I discovered the feather of an eagle given to us months earlier by a Native American in South Dakota. 'It's a symbol of swiftness and courage', I remembered him saying. Seeing it once more revived our spirits and jubilantly we continued on the last leg of our journey down to New Orleans – that city, as William Faulkner wrote, 'created for and by voluptuousness'.

I remembered also, somewhere around Rulo on the lower reaches of the Missouri, what someone had said when I told them of our journey. 'That ain't just any journey', he replied, 'that's what I'd call an *odyssey*.' Perhaps he was right. It certainly felt like one.

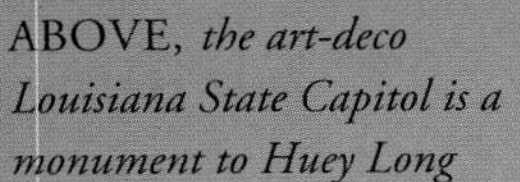

ABOVE, *the art-deco Louisiana State Capitol is a monument to Huey Long*

ABOVE, *Baton Rouge – capital of Louisiana and the USA's fifth largest port*

PRACTICAL INFORMATION

- The Avon 3.4m inflatable dinghy is excellent, but a more powerful engine than 4hp is recommended.

- NW Airlines has daily flights direct to Minneapolis from London Gatwick. Between Minneapolis and Pierre there is no direct bus route and no train service. It is best to negotiate a fare with a cab. The journey to Pierre takes about eight hours by road.

- A descriptive list of maps and charts can be obtained from:

US Army Corps of Engineers
Omaha District, Cemro – OP – N
215 North 17th Street, Omaha, NE 68102 4978. Tel: (402) 221 4175.

- The US Army Corps and Coastguards can provide excellent advice and information on any aspect of boating on the Missouri and the Mississippi.

- Campsites along both rivers are generally good and reasonably priced, but sandbanks are more numerous and cost nothing. Bed and breakfast accommodation can be expensive. Motels, especially Super 8, are cheaper.

- The best time to travel is between July and October.

- Consult your doctor about vaccinations and a medical kit.

LEFT, *New Orleans is a welcome sight at the end of a long and eventful journey*

ABOVE, *fishing paraphernalia becomes a familiar sight on this trip*

BELOW, *St John's, capital of Newfoundland, Canada's easternmost province*

Working the Labrador Coast

ELIZABETH CRUWYS AND BEAU RIFFENBURGH

Northern Ranger **is a not a cruise ship: it is a working supply vessel which provides a vital lifeline to the remote fishing communities along the coast of Atlantic Canada. The journey from the port of St Anthony in Newfoundland to the Inuit village of Nain in Labrador and back again takes about 12 days. Covering a distance of approximately 2,380 miles (3,800km), it meanders through some of the region's most spectacular scenery. Towering icebergs, breaching humpback whales, seabirds by the thousands, plus Viking and Inuit archaeological sites, are just a few of the things that make this a remarkable and memorable journey.**

As folks with more time than money, when we first heard about the Labrador coastal ferry it appealed to us merely as a practical way of travelling from Newfoundland to northern Labrador. In fact, it turned out to be one of the most memorable voyages we have yet experienced. The travel agent said nothing about the pods of humpback whales we would encounter, or about the rugged mountains, towering icebergs, ancient history and vast flocks of seabirds.

The journey started in St John's, the bustling capital city of a province often shrouded in fog. From there, it was necessary to travel by bus across the rocky Avalon Peninsula, past a remote airstrip with the grand name of Gander International Airport, and on through the spectacular Gros Morne National Park to the little port of St Anthony on the Great Northern Peninsula. We arrived at 10pm, and *Northern Ranger* was due to leave at about eight the following morning. Three days and no ship later, the unperturbed harbour clerk finally confided that he thought *Northern Ranger* might be 'a little late'.

TIME IN ST ANTHONY

Northern Ranger is freight carrier first, and cruise ship second, which is why its passengers are treated to sights and experiences most travellers miss. A three-day delay was a small price to pay for the journey that lay ahead. And if you have to take an unexpected vacation, St Anthony during the annual cod festival is not a bad place to be for a day or two. People travel in from miles around and the whole port has the atmosphere of a carnival. Events include not only local music and dance, but demonstrations of skills such as fish-sorting and shrimp-shelling. The final of the cod-filleting competition had every bit as nail-biting a climax as any international sporting event.

Of course, the festival meant that every room in the area had been fully booked for weeks in

St Anthony is a pretty fishing port and the largest town on Newfoundland's Northern Peninsula. The Sir Wilfred Grenfell Mission is located there, founded in 1892 to provide medical services for the scattered people in the area. Today, the Mission is a fine museum, famous for handicrafts such as hand-embroidered parkas.
L'Anse aux Meadows is a UNESCO World Heritage Site and one of the most important archaeological sites in North America. In the 1960s and 1970s it was excavated and found to contain a number of Viking timber and sod buildings, proving that Europeans had discovered North America centuries before Columbus made his epic journey.
The Strait of Belle Isle divides Newfoundland from Labrador and is a turbulent stretch of vivid blue water often infested with icebergs. The entire area from St Anthony to Nain is often referred to as Iceberg Alley because of the huge number of bergs that float there.
Red Bay is a National Historic Site, where a Basque whaling port was founded in the 16th century. Today, visitors can visit the impressive Interpretation Center and discover how this isolated settlement became a world whaling capital.
Happy Valley-Goose Bay is dominated by its airbase, where monstrous Hercules and Ajax transport planes thunder up and down its runways. The annual Canoe Regatta gives the settlement a carnival atmosphere during August, while winter sports are on offer once the snows arrive. There are also a number of museums and galleries, including the Labrador Heritage Museum and the Northern Lights Military Museum.
A Moravian mission church raised in 1782 in Hopedale is now considered to be the oldest wooden building in Atlantic Canada.
Davis Inlet is home to the Innu, people whose nomadic ancestors once hunted caribou in Labrador's vast interior. Further north, in Nain, visitors will meet Inuit, who offer demonstrations of soapstone carving and other traditional skills.

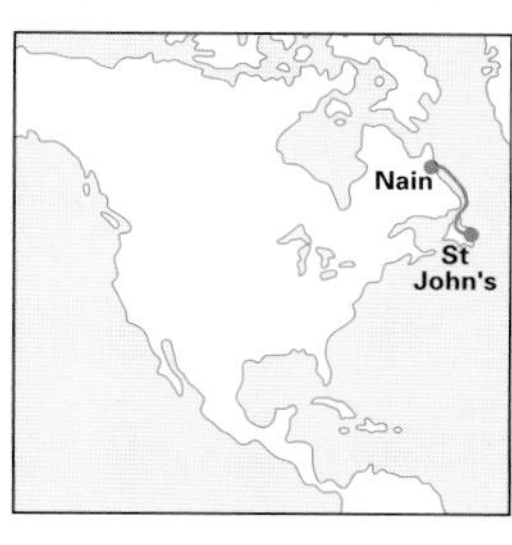

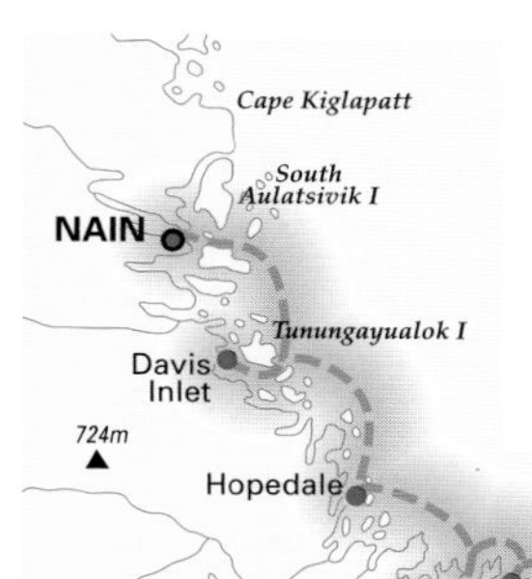

LEFT, colourful paintwork in St John's

ABOVE, *Bonne Bay, a deep fjord, splits the beautiful Gros Morne National Park in two*

BELOW, *a Viking boat at L'Anse aux Meadows National Historic Site*

advance. Fortunately, one cheery fisherman stepped in just as we were wondering what it would be like to sleep under the stars and offered us his room on the grounds that he didn't plan on doing much sleeping as long as the festival was in full swing. The hotel, deserted all night, was packed with exhausted people huddled in sleeping bags on every available scrap of floor from morning until mid-afternoon.

Apart from the festival, St Anthony's main claim to fame is the Sir Wilfred Grenfell Mission, established by Grenfell himself in the 1890s. There is a museum dedicated to the man who spent much of his life raising funds for hospitals, nursing stations and children's homes. Grenfell's house, standing on a wooded hill overlooking the harbour, is also open to the public.

Finally, *Northern Ranger* arrived, a sleek blue and white ship with a yellow trim and spotless orange decks that were constantly being swabbed and cleaned. Virtually all the other passengers were local fishermen who were travelling north for the seasonal work along the Labrador coast. We shared a cabin with two women who were going to visit their husbands in Black Tickle, and who planned to work on the fish-processing plant there. The cabin was soon crammed with their luggage, which comprised such diverse items as a step ladder, three ancient dining-room chairs, two enormous straw sombreros, and a cage of canaries. They also carried enough food to feed an army and so the excellent meals available in the cafeteria were well supplemented with a range of home-cooked cakes and biscuits.

Columbus and Cod

The first port of call was L'Anse aux Meadows National Historic Site, where a bus was provided to take passengers to one of the most significant archaeological sites in North America. The story goes that in 986, a Viking trader called Bjarni Herjolfsson was blown off course while sailing from Iceland to Greenland. When he finally reached Greenland, he reported that he had seen lands to the west and, 15 years later, Leif Eriksson, the son of Eric the Red, set off to find them. He discovered a place he called Vinland and founded a settlement there.

The exact location of the legendary Vinland gave historians cause for speculation for many years. Then, in 1960, the Norwegian archaeologist Helge Ingstad visited L'Anse aux Meadows and excavated some timber and sod huts, noting in growing excitement that they were very similar to sod huts in Scandinavia. Definite proof that L'Anse aux Meadows was indeed a Viking settlement came with the discovery of a bronze, ring-headed cloak pin. The implications are clear: the Norsemen had settled North America 500 years before Christopher Columbus 'discovered' the New World. Today, the site includes an excellent visitor centre, reconstructed timber and sod buildings and offers guided tours.

The next stage of the journey took us across the turbulent Strait of Belle Isle, an 11-mile (18km) wide channel through which the icy Labrador Current races to empty into the Gulf of St Lawrence. It is often littered with ice, even in the summer, ranging from great blue-white icebergs carved by wind and time into extraordinary shapes, to smaller chunks known as 'bergy bits'. The whole stretch of water north to Nain is known as Iceberg Alley and it is a good spot for whale- and bird-watching.

One sunny July afternoon we were treated to a fabulous display of acrobatics from a pod of humpback whales that was migrating north in search of krill and capelin. Few things are as spectacular as a 50-ft (15m) long leviathan surging out of the water just a short distance from the ship to plunge backwards in a great fountain of spray. One whale breached so close that some of the crew took an unexpected shower, and we could smell its distinctive breath. The captain stopped the ship and passengers and crew alike watched spellbound as the display continued for the best part of an hour before these majestic creatures finally continued their journey.

Once the strait had been crossed, we stopped at Red Bay, yet another of North America's important historical sites. In the 16th century, Basque sailors founded a whaling station there. Archaeologists have discovered a galleon, thought to be the *San Juan*, which sank in a storm in 1565; delicate glasses and jugs, and harpoons and whaling equipment. The visitor centre has a highly informative hour-long video about the site.

BELOW, *icebergs are frequently encountered off the coast of Labrador, even in high summer*

Between Red Bay and Cartwright, *Northern Ranger* made about 25 different stops, all at tiny seasonal fishing communities with names like Battle Harbour (settled in the 18th century and beautifully restored), Snug Harbour (tucked under dramatic mountains in an ice-littered cove), and Indian Tickle. Some of these small communities have wharves, but, for the most part, loading and unloading was straight over the side of the ship into frail little boats that appeared in flotillas when *Northern Ranger* sounded her horn. At 5am one morning we saw a vast sofa and two chairs lowered expertly into a sagging dingy leaving barely enough room for the driver. Incredibly, it reached land without capsizing.

Passengers are free to wander whenever *Northern Ranger* docks, although ship's business can be as brief as ten minutes. In most places, ten minutes was about all we needed, given that swarms of persistent blood-sucking blackflies instantly descended on every scrap of unprotected skin, reminding us why the interior of Labrador has never been a popular place for permanent settlements.

Many of the communities face hard times because of the collapse of the fishing industry, and the locals supplement their incomes by selling home-made goods. One elderly fisherman offered Liz his hand in marriage, but still presented her with a huge cod as a gesture of goodwill when she declined. She was flattered until the ship's cook informed her that wives were in short supply in this part of Labrador, and that lonely men weren't particularly fussy about whom they asked – as long as she could gut a cod. The cod was turned into a Newfoundland dish called 'fish and brewis' (pronounced 'brews'), a sort of hardtack and cod mash that looks like soggy cardboard, but tastes surprisingly good. The crew members, who ended up taking a huge proportion of it, informed us it was one of the finest dishes this side of Mars.

One of the officers on *Northern Ranger* was a local folk singer of some repute and he was under constant pressure to provide impromptu concerts along the way. He declined, but the captain played one of his tapes over the loudspeaker system, thus treating the entire ship to an hour of lively Newfoundland reels. It was just one example of the friendly, relaxed attitude of the crew and their easy relationship with the people for whom the ship is an essential part of life.

BELOW, *a breaching humpback whale – one of the exciting sights of the trip*

MISSIONS AND TRADITIONS

After Cartwright, which was named after a merchant adventurer who lived in the area in the 18th century, *Northern Ranger* cut inland for 130 miles (208km) along Hamilton Inlet to Lake Melville. The lake was silent and still, disturbed only by loons that slapped across its glassy surface. Thickly wooded shores gave way to the dramatic Mealy Mountains in the distance.

At the head of the lake is Happy Valley–Goose Bay, a bustling community with a frontier atmosphere and a large military airbase. Visitors can take the Trans-Labrador Highway (Route 500) to the hydro-electric facility at Churchill Falls, past the spot where Leonidas Hubbard starved to death in 1903 – despite eating his moccasins – when he tried to explore this inhospitable terrain. It was left to his remarkable wife Mina to complete his work when, two years later, she returned to Labrador, walking across 550 miles (880km) of barren wilderness in just 61 days.

Retracing her route, *Northern Ranger* glided back towards the open sea and turned north. At the head of Kaipokok Bay is Postville, an old fur-trading post that was established in 1843. But Postville boasts a much more ancient history, and there is evidence that people of the Dorset Eskimo culture came here in the summer to fish and hunt some 4,000 years ago.

Labrador's close association with the Moravians can be seen at Hopedale, which boasts both the Moravian Mission Museum and the oldest wooden structure in Atlantic Canada. At Davis Inlet, visitors can buy goods made locally by the Innu (once called Naskapi Indians), people whose ancestors eked a precarious existence from the land and the sea. Traditional crafts and skills are still practised, despite the people having been relocated from tents to houses in the 1960s.

Northern Ranger's final stop is at Nain, a remote settlement with unpaved roads and a jumble of wooden houses. It is the administrative centre of the region's Inuit and guided tours are available to demonstrate how these people made their living in the north. The charmingly named Piulimatsivik (an Inuit word meaning 'place where we keep the old things') houses a fascinating collection of artefacts from both aspects of the community's past – Inuit kayaks and hunting

equipment, and a history of the Moravian mission that was established in 1771.

And so the journey ended. Some of us disembarked in Nain to travel farther north to see the abandoned mission at Hebron and the wild Torngat Mountains, one of the world's great attractions for mountaineers. Other passengers immediately rejoined *Northern Ranger* for the run back to St Anthony. Horn bellowing, she sailed away, her hold stuffed with goods to be delivered along the way – a piano for an anxiously waiting man in Cartwright, cases of beer and new light-bulbs for the seasonal fishermen at Ice Tickle, and a big stuffed animal for a new baby in Rigolet.

The itinerary is unpredictable on *Northern Ranger*, depending on weather, workload and the tides, but it can guarantee awe-inspiring scenery, icebergs, whales and as much traditional Newfoundland and Labrador hospitality as the visitor can absorb.

LEFT, *a scene typical of life along the coast of Newfoundland*

PRACTICAL INFORMATION

■ Coastal cruises in Labrador are run by Marine Atlantic (Marine Atlantique), which has offices in Port aux Basques, Newfoundland (Tel: 709 695 7081) and North Sydney, Nova Scotia (Tel: 902 794 5700) in Canada, and Bar Harbor, Maine (Tel: 800 341 7981) in the United States.

■ *Northern Ranger* can carry up to 131 passengers, although there are only 83 berths in 23 cabins. Advance booking is necessary. Travel packages include accommodation, cafeteria-style meals, and guided tours at some places. Travellers usually arrange their own transport to St Anthony.

■ For passengers with less time, eight-day cruises are available in June only, travelling from St Anthony to Goose Bay and back.

■ Cruises leave every two weeks or so between July and November.

■ For passengers travelling in the high summer, insect repellent is a necessity, not an optional extra! Warm clothes, sturdy shoes and rain-gear are also recommended. If you forget something, only St Anthony and Goose Bay have shops.

ABOVE, *souvenirs laid out to catch the eye of passengers disembarking on St Martin*

Stepping out in the Caribbean

SHIRLEY LINDE

This is a traditional voyage aboard a grand old ship, a trip recalling the stylish, bygone days of Atlantic cruising. For lovers of swing music and ballroom dancing from the Big Band era it is a must, with nostalgia and dancing partners in abundance. Leaving Miami on Saturdays, the SS *Norway* spends the first two days at sea before visiting some of the best-known islands of the Caribbean, finally fetching up at Great Stirrup Cay in the Bahamas. From start to finish, the beat doesn't waver.

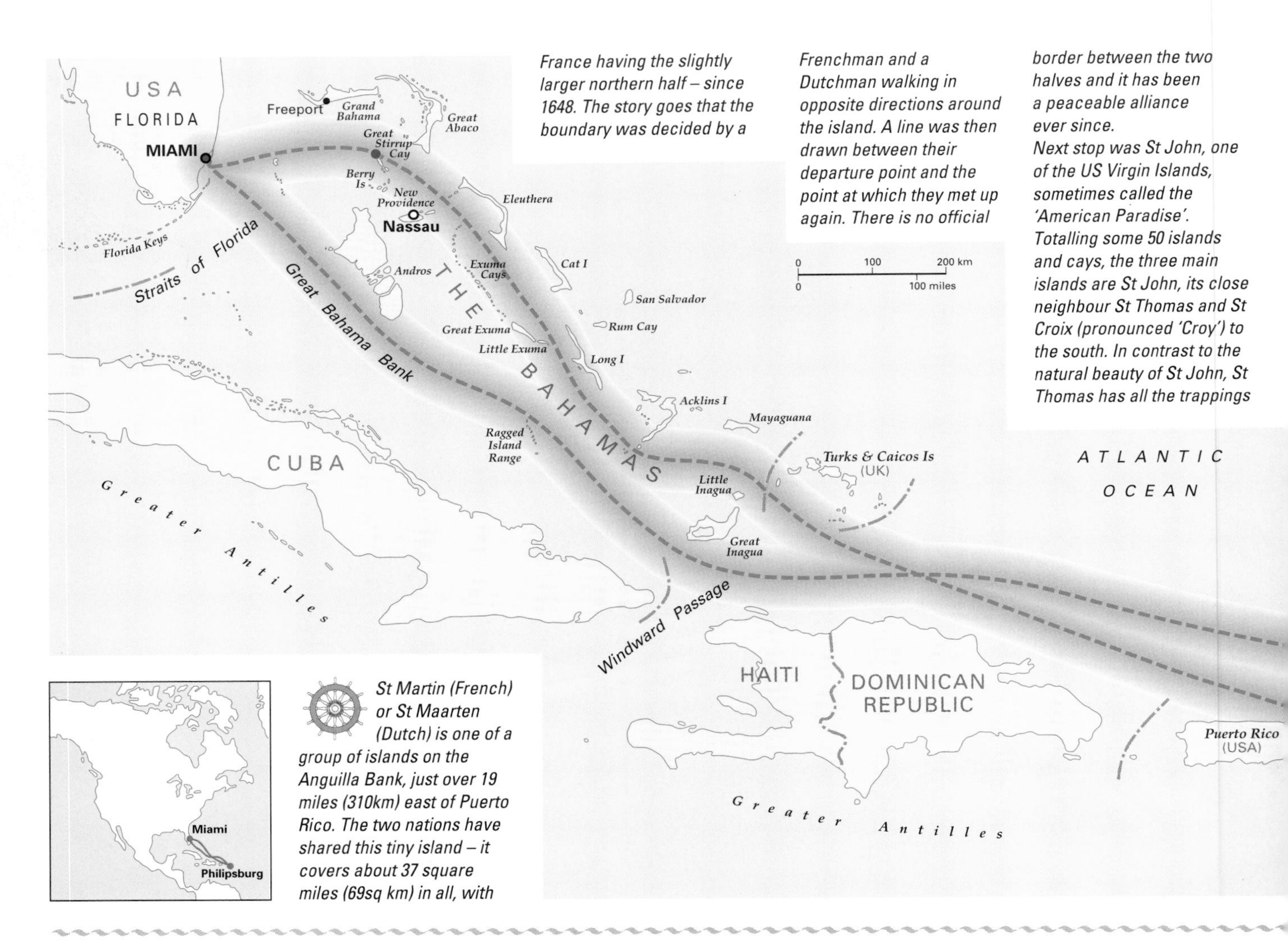

St Martin (French) or St Maarten (Dutch) is one of a group of islands on the Anguilla Bank, just over 19 miles (310km) east of Puerto Rico. The two nations have shared this tiny island – it covers about 37 square miles (69sq km) in all, with France having the slightly larger northern half – since 1648. The story goes that the boundary was decided by a Frenchman and a Dutchman walking in opposite directions around the island. A line was then drawn between their departure point and the point at which they met up again. There is no official border between the two halves and it has been a peaceable alliance ever since.

Next stop was St John, one of the US Virgin Islands, sometimes called the 'American Paradise'. Totalling some 50 islands and cays, the three main islands are St John, its close neighbour St Thomas and St Croix (pronounced 'Croy') to the south. In contrast to the natural beauty of St John, St Thomas has all the trappings

I live alone, in America, so sometimes traditional family holidays such as Thanksgiving and Christmas can be lonely. Sometimes, on the other hand, they turn out to be nothing but hassle – running errands through snarls of traffic, shopping amid impatient crowds, unpacking and repacking decorations, standing over the stove for hours cooking for a crowd.

This year, however, I found the answer – instead of being alone *or* hassled, I took a cruise. Once you accept that holidays will get along fine without you all sense of guilt will vanish and you'll feel a wonderful sense of freedom. Nothing to do but pack!

There are very many cruises to choose from but I picked the SS *Norway,* of the Norwegian Cruise Line, not least because she is a grand old ship, the kind I had always wanted to travel on. This particular seven-day cruise spanned Thanksgiving and left from Miami to cruise the Caribbean. Called a 'Big Band Cruise', it came complete with dance hosts and four well-known swing bands from the past. Perfect – I like jazz and big band music and I love to dance, but often don't have a good dance partner. Without further ado I signed up.

A FLOATING PALACE

The SS *Norway* is the flagship of the Norwegian Cruise Line fleet. Christened the SS *France* on 11 May, 1960 by Madame Yvonne de Gaulle, at 1,035ft (315m) she is the longest passenger ship ever built. Along with Cunard's *Queen Mary* and *Queen Elizabeth*, she was one of the great luxury ships that regularly crossed the ocean, a floating palace enjoyed by a galaxy of royalty, celebrities and dignitaries.

When passenger cruises became unprofitable (air travel was quicker and cheaper) sailings were stopped and from 1974 the *France* languished in the port of Le Havre. Eventually, pioneer of the cruise industry Knut U Kloster bought her for $18 million for his Norwegian Caribbean Line (today known as the Norwegian Cruise Line), and renamed her the SS *Norway* in honour of his homeland. Over the next ten months some 2,000 workers renovated the ship at a cost of $100 million. The *Norway* began her new life in the Caribbean, sailing on her first seven-day cruise on 1 June, 1980. She left from Miami, her new home port, with an international crew of 800 drawn from some 40 nations.

ABOVE, *exotic sealife such as this queen angel fish can be seen by taking one of the many snorkelling tours available on St John*

of tourism and is one of the most developed of the Caribbean islands. Central to its history is the splendid harbour, and the 18th-century warehouses in Charlotte Amalie are testimony to its days as a trading port.
Some 700 islands plus about 2,000 cays scattered over the Atlantic Ocean make up the Bahamas, synonymous with blue seas, beautiful beaches, superb diving and easy living. During the last 40-odd years these low-lying coral outcrops have become a majour tourist destination with an estimated 3 million visitors a year, about half of whom arrive on cruise ships such as ours.

BELOW, *Miami's skyline disappearing behind us as we set off on our journey*

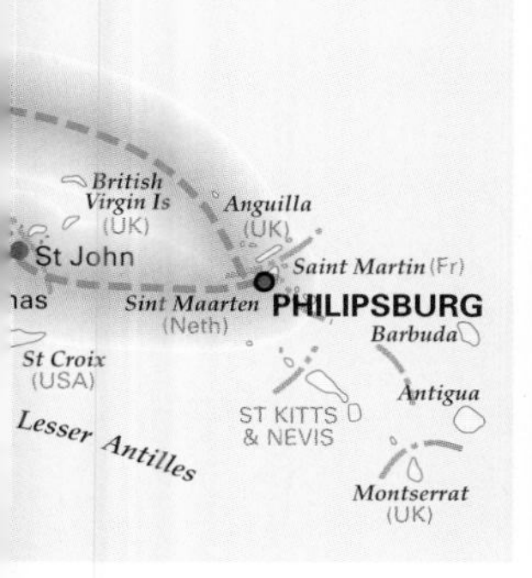

MAIN PICTURE,
SS Norway *approaching the harbour of Charlotte Amalie on the US Virgin Island of St Thomas*

INSET OPPOSITE,
Marigot Harbour, on the French side of St Martin, viewed from Fort Louis

The *Norway* maintains a style and graciousness reminiscent of an earlier cruising era. She has been refurbished several times, but the art deco murals, hand-laid tile mosaics, polished teak rails and nautical antiques acquired from bygone cruises have been carefully preserved, and when walking along the long promenade deck there is a feeling of stepping back in time. In the restaurant, you are aware of sitting where famous stars have dined, and later that evening one half expects Cary Grant, elegant in his tuxedo, to stroll down the deck and lean nonchalantly next to you on the rail.

It takes a while to find your way around as the *Norway* is 10 blocks long and 12 decks high. There is an International Deck lined with sidewalk cafés and boutiques; a fitness centre with glass walls so you can exercise while gazing out at the sea on the Olympic Deck; on a lower deck you come upon a decadent Roman spa that offers massages, aromatherapy, body wraps, saunas, steam rooms, a jacuzzi and an indoor pool for water exercise. There are two outside pools, and the Olympic Deck has a jogging track. In addition, there are seven bars, six entertainment lounges, one grand ballroom, a cabaret, a large casino, a disco, an ice-cream parlour, a library, a piano bar and a cinema.

The *Norway* has 1,039 staterooms in all, each with individually controlled air-conditioning, private bath with shower, television, radio and phone. Some suites have a separate living room and bedroom in addition to a master bedroom. Most of the luxury penthouse suites also have private balconies.

One evening a number of officers and passengers, myself included, were invited to a party in what had to be one of the fanciest of the penthouse suites. The accomodation was spectacular, with a wrap-around balcony, living room and bedroom, dressing room and jacuzzi. Our host had won a lottery in his state of several million and was celebrating in style.

ENTERTAINMENT ON TAP

The hardest task each morning was choosing what to do that day. Options included Broadway shows, exercise classes, dance instruction, basketball, golf driving and putting, paddleball, ping pong, shuffleboard, skeet shooting, snorkelling classes and excursions; volleyball, fashion shows, wine-tasting, art auctions, lectures. There weren't many youngsters aboard this cruise, but usually there is a youth programme with a children's playroom, activities for all ages and special shore excursions – not forgetting a tea for grandparents, a 'mixer' party for single people, and a champagne party for honeymooners. How can anybody think a cruise might be boring?

We left Miami late in the afternoon and the activities started immediately. Some passengers headed for the casino, waiting for it to open when we got outside the legal limit. It was set up for blackjack, craps, roulette and baccarat plus 200 or so slot machines. I was happy to stay on deck, listening to the music, snacking off the welcome-aboard buffet and watching the shoreline disappear astern.

As I watched the waves break alongside the ship and the wake trail behind, I was already glad that I had made the decision to come away: no holiday rush, fresh air and no traffic jams – unless you count the jostle as people flocked into the dining room to dig in to yet another meal. The only decisions I had to make were whether to shop on board or in port, or both, play golf or tennis, go swimming at some tranquil tropical beach or work hard at deep-sea fishing. At night, more difficult choices had to be made between delicacies such as conch fritters, shrimp, fresh baked bread and blueberry muffins; stuffed Cornish hen, grilled swordfish and coconut meringue pie, or that 'death by chocolate' dessert.

ABOVE, *Trunk Bay, one of the unspoilt beaches on the northern shore of St John, belonging to the national park*

BELOW, *dancing on a little less formal scale*

DANCING THE NIGHT AWAY

By the first night passengers were getting to know each other and serious dancing was already underway. I had never met such a concentration of people who enjoyed tripping the light fantastic so much and who knew so much about jazz and big band music. I wandered from ballroom to ballroom, or sat on the International Deck talking to other passengers reminiscing and comparing memories. In the background a complementary 24-hour CD jukebox played choice selections of jazz and big band recordings.

Four bands were playing that week: the Tommy Dorsey Orchestra conducted by Buddy Morrow, Si Zentner and his Orchestra, the Bob Crosby Orchestra conducted by Ed Metz Junior, and the Harry James Orchestra conducted by Art Depew. The Café International, which turned out to be my favourite place, featured a nightly band whose drummer used to play the background music for the old Fred Astaire movies.

I met the four gentlemen dance hosts immaculate in their navy blazers and white slacks and danced my first dances.

About seven years ago an organisation called Ballroom Dancers Without Partners was formed and this was represented on board. The group is made up of men and women from around the world who like to dance but who lack partners. Members attended a get-acquainted cocktail party, private dance instruction was available if you so desired, and there was dancing every night in various ballrooms until the small hours. Their four hosts were in addition to the Norwegian Cruise Line's six dance hosts, and many of the men belonging to the group turned out to be excellent dancers.

A day or two after setting sail the staff began to put up and decorate Christmas trees around the decks and the video channel featured holiday films in addition to the period movies, big band performances and interviews with well-known big band stars that had been featured all week. Workshops with the band leaders were also held mid-week. We were all happy to enter into the holiday spirit with ease, as here it was different, there was no pressure.

ISLAND HOPPING

Our first port stop was St Maarten (St Martin), a tiny island actually shared by two nations: the southern half is Dutch and the northern half is

French. We docked at Philipsburg, the Dutch capital and the island's main port where most cruise ships moor. Typically West Indian, its four main streets are lined with pretty pastel-coloured buildings decorated with gingerbread fretwork. Just the other side of the imaginary dividing line is the French capital, Marigot, with its pavement cafés, somewhat faded colonial buildings and attractive waterfront with bistros and boutiques. Both sides of the island have many good beaches to suit all tastes. Duty-free shopping is available everywhere too, with a choice of inexpensive T-shirts, or more expensive jewellery and French and Caribbean designer clothing.

In the morning I chose to go on an exciting excursion aboard a 39ft (12m) racing sailboat that raced in the America's Cup. Later I wandered from shop to shop in the sun while calypso music played in the background, then stopped for a cool drink on a restaurant patio overlooking the ocean, and thought about the people shopping in the crowds back home.

Our next stop was St John, one of the the US Virgin Islands. Two-thirds of the island is designated a protected national park so it is quite unspoilt with empty beaches and a number of excellent hiking trails. I chose to go sailing again, while others opted for lying on the beach, sightseeing around the island by safari bus, or going on one of several snorkel/scuba dives. Our sailboat took us over to St Thomas, where we caught up with the ship. Many passengers went into town to do some duty-free shopping – there are some people you just *have* to buy Christmas presents for, while others headed for Magens Bay, arguably the island's most fabulous beach and certainly very popular. Another group went snorkelling at the protected reefs of Buck Island, where feeding fish by hand underwater is a highlight, and others viewed coral and sea life from an Atlantis submarine. We were all back on board as usual in plenty of time for a rest and a shower before dinner, and of course, more big band music and dancing.

The next day was spent on the beach on a little island in the Bahamas – Great Stirrup Cay. I lived in the Bahamas for five years on one of the out islands, so it was wonderful to get back to those transparent turquoise waters.

At the end of the seven days it was hard to believe that the cruise was over already. By now, the ship was sparkling with decorations, ready for the Christmas and New Year's cruises. They were already totally booked but, I thought, I can sign up for next year; in fact I learned that the Big Band cruise was so popular that the Norwegian Cruise Line was going to have another in the spring in addition to the traditional Thanksgiving trip. This would feature the Woody Herman Orchestra, the Artie Shaw Orchestra, Larry Elgart and Hal McIntyre. And then there are the Norway's Country and Western cruises and the Jazz cruises, and the other cruises of the Ballroom Dancers Without Partners Association on *Holland America*, and some small-ship cruises such as the Clipper Cruise to Costa Rica and Panama and the Marine Expeditions Cruise to the Antartica. I don't think holidays are going to be a problem any more.

PRACTICAL INFORMATION

- United States and Canadian citizens must have proof of citizenship and photo identification. Others need a valid passport and a multiple-entry visitor's visa.
- Take clothes for two formal nights (cocktail dresses or gowns; tuxedo or jacket and tie), two informal nights, one country and western night, one Caribbean night and one 50s/60s night. You will need a sweater or jacket for evening wear as the lounges are often too cold, but you heat up fast after a few cha-chas and swings.
- On choosing your cabin, note that some of the less expensive ones on the inside are very small. Ten cabins are designed for wheelchair passengers. Portside cabins are non-smoking.
- Safety-deposit boxes are available at no charge.
- When you book, be sure to tell your travel agent or the cruise line if you are celebrating a birthday, anniversary or family reunion, or if you are on your honeymoon. You should also let it be known ahead of time if you have any special dietary needs.
- When arranging return air tickets, bear in mind that disembarkation usually doesn't begin until about two hours after the 8am docking in Miami.
- For further information on *SS Norway* cruises, call Norwegian Cruise Line, Tel: 800 327 7030 or 305 436 0866 in Florida or contact your local travel agent.

LEFT, *Magens Bay, one of the nicest and most popular beaches on the island of St Thomas*

Atlantic to Pacific through the Panama Canal

GARY BUCHANAN

ABOVE, *pampered passengers cruise through the Panama Canal in style*

There cannot be a more splendid way of crossing the continent of North America than aboard the *Crystal Symphony*, a floating haven of great luxury and wealth. The journey is made possible by the multi-lock waterway of the Panama Canal which runs across the mountainous terrain of the Isthmus of Panama for 51 miles (82km), representing an unprecedented feat of engineering. As she makes her stately way from ocean to ocean, the ship puts in at Grand Cayman, Aruba and Acapulco, places where the romance and glamour match that found aboard.

BELOW, *situated between Miami and Palm Beach on Florida's Gold Coast, Fort Lauderdale provides a mix of culture, history and sports*

At 5pm, Captain Helge Brudvik orders the gangway to be taken in. 'All clear fore and aft! Cast off all lines!' With a gentle rumble *Crystal Symphony* comes to life, imperceptibly at first then as the bow thrusters gather momentum the shimmering white hull eases away from her pier at Port Everglades. An evening shadow reflects the ship's stately progress across the terminal roof for a few fleeting moments, then with a graceful list to starboard we head out to sea. The north shore of the channel leading from the port of Fort Lauderdale is lined with condominiums standing sentinel in a salute to the passage of a myriad of cruise ships as they begin their passage out to sea. With a light breeze tempering the heady air enveloping the warm Atlantic Ocean, one of the world's most luxurious vessels heads south to begin her transcontinental passage to the Pacific Ocean. As the metropolis of Miami appears on our starboard bow, the slowly sinking sun illuminates the pastel-hued art deco promenade of South Beach, creating a surreal silhouette.

NO EXPENSE SPARED

From the moment we arrive at the two-deck high Crystal Plaza atrium, complete with spiralling waterfall backdrop, we're bowled along on a velvet-lined conveyor belt attended by a flotilla of flunkies proffering silver trays laden with champagne bubbling in Lalique flutes. A brief orientation tour reveals a wealth of intimate hideaways such as the cocktail-friendly Avenue Saloon,

LEFT, *elegant yachts line the marina at Fort Lauderdale at the start of the journey*

On this trip, the Crystal Symphony set off from Fort Lauderdale in Florida, cruising round the northern tip of Cuba to reach the Cayman Islands. The largest of the trio, Grand Cayman, is 22 miles (35km) long and 8 miles (13km) wide, flat and shaped like a potato with a large hook at the western end. A number of small communities dot the coast, while stretching along the seaward edge of the hook is Seven-Mile Beach, a classic strip of pure white sand with a terrace of a dozen or more hotels, condominiums and luxurious private estates.

The enchanting cornucopia that comprises the Dutch Antilles of Aruba, Bonaire and Curaçao lies off the northern coast of Venezuela. This is about as disparate a trio of islands as it is possible to imagine, with each island having a distinct personality of its own.

The creation of the Panama Canal represented the largest, most costly single undertaking ever before mounted anywhere on earth, and it held the world's attention over a span of 40 years. The cost of construction was $387 million, but the price paid in human terms, as a result of tropical diseases and heat exhaustion, was estimated to be 26,000 lives. The first journey along the canal was made on 15 August 1914 by the SS Ancon, *and at a stroke the arduous 8,000 mile (12,800km) voyage around Cape Horn that could take up to several weeks, or even months, was reduced to a mere eight or nine hours.*

There are three sets of locks, each chamber measuring 1,000ft (304m) in length, 110ft (33m) in width and 41ft (12.5m) in depth. The lock gates are so finely balanced that, despite their enormous weight, they can be moved by a 40hp motor.

At the Gatun Locks, vessels are lifted in three stages to the canal's highest point, the Gatun Lake, 85ft (26m) above sea level. Electric locomotives, known as 'mules', running on tracks on both sides of the locks haul ships through on massive steel towing wires which also hold them centrally in the lock.

At the Pedro Miguel Lock, vessels are lowered 31ft (9.5m) to the Miraflores Lake. Thirty minutes or so later they enter the Miraflores Locks, which lower ships 54ft (16.5m) down to the Pacific sea level; this can often be up to 2ft (0.5m) higher than the sea level of the Atlantic Ocean.

On 31 December 1999, as required by the Panama Canal Treaty, the United States will transfer ownership and operational control of the canal to Panama.

Three more days easy travelling takes the ship to Acapulco, acclaimed – and rightly so – as Mexico's leading tourist resort.

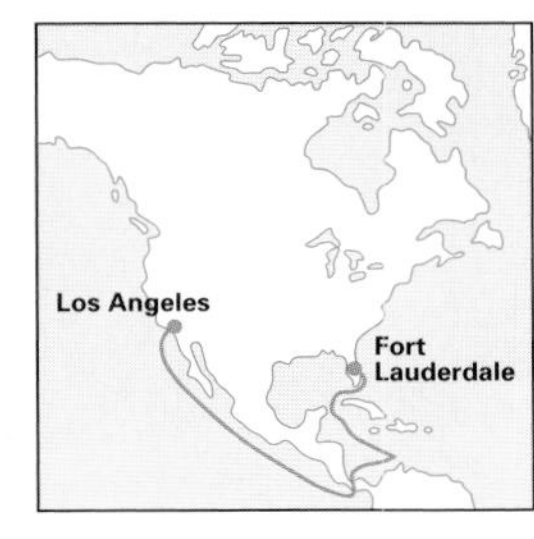

CRYSTAL SYMPHONY

ABOVE, *elegant surroundings, fine food and impeccable service characterise dining on board ship*

the French-style Bistro café and the cosy rendezvous of the Crystal Cove – complete with glass-encased piano. It's summertime on the *Crystal Symphony*, and the living is easy.

At 50,202 tons, the ship could comfortably accommodate almost 2,000 passengers but, in keeping with the line's philosophy of all-consuming luxury, only 960 are conveyed in this palace of the sea. Even then, we're scarcely conscious of even half that number since the variety of lounges, restaurants and bars creates the impression of a more intimate vessel.

In time-honoured tradition the first night at sea is devoted to ball gowns and black ties. The Starlite Club positively drips in designer frocks, while the spotlights reflect a kaleidoscope of precious stones. As one inveterate lady of the seas observes, 'Where else in the world can I wear all these wonderful stones my husband bought me and still feel completely safe?' Captain Brudvik and his senior officers look resplendent in their white mess uniforms and after a brief introduction and good wishes for a 'bon voyage', the personable cruise director invites us to take our seats for the Captain's Gala Dinner.

TURTLES AND WINDMILLS

Having spent a leisurely day navigating around the north-western tip of Cuba, we rise early the following morning; as the dawn breaks, *Crystal Symphony* makes landfall on Grand Cayman, largest of the Cayman Islands. From our veranda we can see swathes of Australian pines fringing the endless beaches that are turning a pale yellow.

At the end of the island is the world's only turtle farm, constructed in 1970 as a biological research station; it is now a tourist attraction. Centuries ago these uninhabited islands were a favourite spawning ground for turtles, but following their discovery by Columbus in 1503 – who named the island group 'Las Tortugas' – these docile reptiles were hunted to extinction by pirates and traders. Today the conservation-minded Caymanians have not only adopted the turtle as their national emblem, they have chosen to inflict the ignominy of dressing it up as a pirate on tourist brochures.

George Town, Grand Cayman's capital, has one radio station, one traffic light and over 440 banks. The Caymans' status as a British Crown Colony and tax haven is prized by offshore companies and corporations seeking a little fiscal *laissez-faire* and as a result these three small islands (total population 26,450) are among the most prosperous in the Caribbean.

For the *Crystal Symphony's* wealthy contingent of day-trippers, more concerned at the moment with beaches and coral reefs than banks and tax relief, the day is spent enjoying the perfection of Seven-Mile Beach or, more energetically, under the surface of the tranquil blue sea, scuba diving. With an unprecedented clarity of water ranging up to 200ft (61m), not to mention four large shipwrecks, the Caymans offer some of the very best diving in the world.

OPPOSITE PAGE, *wherever we moored,* Crystal Symphony *dwarfed most other vessels*

INSET, *Oranjestad, Aruba's capital and main commercial port*

Back on board many sun-kissed revellers head for the nocturnal pursuits on offer in Caesar's Palace at Sea. The Las Vegas-style casino reverberates with loud cries emanating from the crap table while the rattling of dollar coins into the hoppers of the slot machines intimates that lady luck is hitting the right note. Marnie from Honolulu isn't interested in any of the ports of call, she could be anywhere on the high seas as long as Caesar's door is open and three lucky sevens choose to appear on the pay line with uncanny regularity.

Another day of languor and indolence passes before *Crystal Symphony* arrives at the port of Oranjestad on Aruba. Despite its size – it is just 20 miles (32km) long and 6 miles (10km) wide – the island offers plenty of scenic variety, the placid waters off the south coast contrasting sharply with those of the north, where the land drops sharply into a turbulent sea. Its highest peak is a 541ft (165m) cone-shaped hill known as the Haystack.

If we are in any doubt about this island's colonial history, Wilhelminastraat, the old quarter of

the quaint capital of Oranjestad, quickly reaffirms this Caribbean outpost's strong Dutch links. Tall, narrow-gabled, 18th-century houses with red-tiled roofs painted in cool pastel shades juxtaposed with intense yellows and royal blues conjure up images of a fairytale Amsterdam, while the wharf and old schooner harbour rekindle scenes of a timeless Dutch port. We lunch in an authentic Dutch windmill, while others opt for an equally authentic *rijstafel* at Bali – an exotically decorated houseboat restaurant.

MAIN PICTURE, *Grand Cayman's Seven-Mile Beach, offering palms, acres of powdery white sand and turquoise waters*

PRACTICAL INFORMATION

■ Crystal Cruises was formed in 1988 as a subsidiary of the NYK Line, the largest shipping company in the world, and throughout the year its two ships, *Crystal Symphony* and *Crystal Harmony*, undertake a series of trans-canal voyages.

■ Sailing from either the Pacific ports of Los Angeles or Acapulco, heading eastbound through the Panama Canal, or westbound from the Caribbean ports of Fort Lauderdale, St Maarten (St Martin), San Juan (Puerto Rico) or New Orleans, they represent the height of luxury cruising.

■ Fort Lauderdale is a most accessible port to join a cruise, while the port of San Pedro is convenient for Los Angeles International Airport. Crystal Cruises schedule their trans-canal sailings to avoid hurricane season in the Caribbean and visit ports that are generally off the itineraries of larger,

THROUGH THE CANAL

As the temperature rises the following day, so does the expectation of a passage through one of the great engineering wonders of the world – the Panama Canal.

ABOVE, *waiting patiently in the middle chamber of the Panama's three Gatun Locks*

BELOW, *Acapulco Bay and its high-rise shoreline, a mecca for sun-seekers and sports lovers*

We opt for an 'early riser' breakfast on our veranda in order not to miss a single moment of the highlight of our cruise. Approaching the port of Cristóbal at the Caribbean entrance to the canal we gently nudge past a flotilla of merchant ships also awaiting their turn to transverse the North American continent at its narrowest point. At 6am we arrive at the Colón Breakwater where the Panama Canal pilot boards – this is the only occasion when the captain of a ship relinquishes his command in favour of the canal pilot – and by 7am the 'crossing of the continent' gets underway in earnest as *Crystal Symphony* enters the three Gatun Locks.

The mid-point of the transit, Gatun Lake – a huge man-made lake created by flooding a valley behind the Gatun Dam – forms the principal source of water for the canal.

All around us is lush, steamy tropical jungle. The thick canopy of trees are ablaze with brightly coloured birds, while the muddy shoreline is home to that most hideous creature – the alligator. After passing Barro Colorado Island – a wildlife preserve of the Smithsonian Institute – we head for Gamboa where we begin our passage through the spectacular Gaillard Cut, an 8-mile (13km) slash through the solid rock of the Continental Divide.

Four miles (6km) after clearing the Miraflores Locks, *Crystal Symphony* passes underneath the Puente de las Americas – Bridge of the Americas – heralding our arrival at the port of Balboa, with the high-rises of Panama City lying off on our

port bow. This impressive steel bridge, as part of the Pan-American Highway, constitutes the only road link between North and South America.

PURE ROMANCE

The next three days at sea are almost an anti-climax. The searing heat and stifling humidity of Panama are replaced by cooling sea breezes as we head north past the coastlines of Costa Rica, Nicaragua, El Salvador, Guatemala and Mexico.

Acapulco is everything tourist brochures and movies had led us to expect – and more. The breathtaking halfmoon bay that spreads out in front of the bow is made even more dramatic by the imposing Sierra Madre del Sur range that creates a spectacular backdrop.

We head for the beach at Caleta, protected from the open sea by the nearby island of La Roqueta and mercifully devoid of *mariachi* bands, to enjoy *pescado à la talla* (barbecued fish) under the shade of a panoply of bougainvillaea. Enterprising local boys, led by a svelte 16-year-old named Francisco, are persuasive in their attempts to sell us straw bags and stuffed armadillos. They insist on taking us to La Quebrada, a place of dramatic beauty facing out to the open sea from which a handful of daring young men – Francisco included – plunge 132ft (40m) into a crevice, cheating death by anticipating the exact moment when the incoming waves will flood it. Never before had we witnessed such a feat, or one more deserving of a handsome tip.

With her passengers all safely back on board, *Crystal Symphony* gives three long blasts on her horn to bid farewell to Mexico's most romantic resort, which by nightfall has transformed into a scene of indescribable beauty with the neon-clad horseshoe of La Costera (the main coastal avenue) fringed with a mantle of twinkling lights .

During the next three days *Crystal Symphony* continues her passage northwards, crossing the Tropic of Cancer at the southern tip of the Baja California Peninsula, *en route* to Los Angeles.

Shortly after daybreak we sail past the *Queen Mary* – that former icon of the heyday of sea travel (she was the flagship of the Cunard Line for some 30 years) now reposing in her own private lagoon in Long Beach – before *Crystal Symphony* arrives at the cruise terminal at San Pedro, journey's end.

After 14 nights as an indulgent retreat, our spacious penthouse has assumed a homely feel and it is with great sadness that we reluctantly pack – declining the offer from our butler to relieve us of this most depressing task.

We have traversed the great American continent in the epitome of luxury and amassed a wealth of good memories. Bidding a fond farewell to our fellow passengers before bundling into a taxi, we catch a glimpse of a white stretch limousine, complete with liveried chauffeur. Neither of us are surprised when it is Marnie's face that appears at the window; she gives us a cheeky wave before raising the tinted-glass and heads off in consummate luxury, thanks, no doubt, to the benevolence of Caesar.

mass-market cruise ships. During the rest of the year, *Crystal Symphony* and *Crystal Harmony* visit Hawaii, Alaska, New England, the Orient, South Pacific, Africa, Europe and South America. Every year, *Crystal Symphony* embarks on a world cruise.

■ Crystal Cruises can be booked through the British sales office at Quadrant House, 80–82 Regent Street, London. W1R 6JB, Tel: 0171 287 9040, in the USA at 2121 Avenue of the Stars, Los Angeles, CA 90067, Tel: 310 785 9300 or in Australia from Wiltrans, Sydney, Tel: 2 9255 0899.

■ Cruises range from ten to 14 days with many of these sailings calling additionally at Playa del Carmen, Cozumel, Puerto Quetzal and Caldera in Mexico.

■ The 960-passenger *Crystal Symphony* features the largest penthouses afloat, the highest stateroom and penthouse/veranda ratio and the leading guest/crew ratio in her class.

■ The Crystal Dining Room operates a two-sitting policy, but this rarely causes any problems as most American passengers prefer an early dinner, while European guests choose the later option. Some regular travellers often forego this lavishly appointed room, preferring instead the intimacy of the two speciality restaurants: the Jade Garden, serving Chinese delicacies, and Prego, offering fine Italian cuisine in a truly cosmopolitan atmosphere.

LEFT, *waves crash against the rocks at Cabo San Lucas at the tip of the Baja Peninsula*

Venturing up the Amazon: a Vanishing World

BEN DAVIES

ABOVE, *daily working life on the river*

BELOW, *the Ver-o-Peso market in Belém, selling produce of all kinds, was originally a slave market*

To travel from Belém at the mouth of the Amazon (Amazonas) River to Tefé in the heart of the Brazilian jungle, a distance of some 1,000 miles (1,600km), is to experience the mightiest river in the world and view a myriad of flora and fauna – not all of it welcome! Most of all, though, it is to witness the fate of the last great rain forest on earth, and not even unprecedented discomforts such as rickety boats, malarial mosquitoes and snapping piranhas can undermine this great adventure.

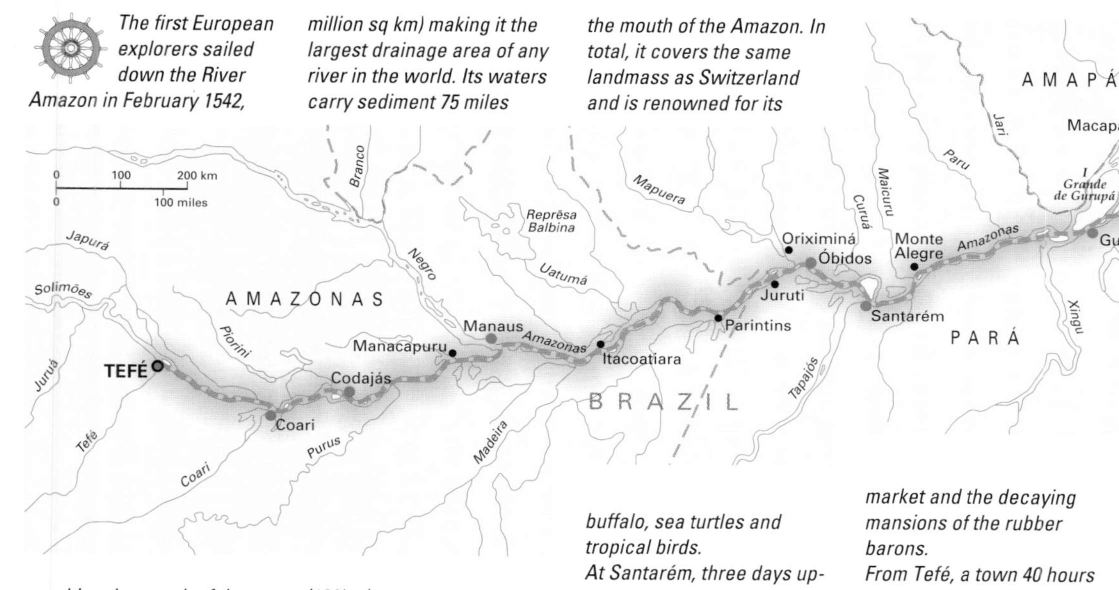

The first European explorers sailed down the River Amazon in February 1542, reaching the mouth of the river some five months later. In total the Amazon (Amazonas) is fed by 1,100 rivers. Its major tributaries include the Japura, the Madeira, the Negro, the Purus and the Xingu, all of which extend more than 1,000 miles (1,600km). The Amazon Basin itself covers 2.7 million square miles (7 million sq km) making it the largest drainage area of any river in the world. Its waters carry sediment 75 miles (120km) out to sea.

In 1616, the Portuguese built the Forte do Castelo near the entrance to the river at Belém. Other sights to see include the colourful Ver-o-Peso market, the 17th-century church of Nossa Senhora das Merces and the splendid Teatro da Paz. Near the city, a huge island called Marajó blocks the mouth of the Amazon. In total, it covers the same landmass as Switzerland and is renowned for its buffalo, sea turtles and tropical birds.

At Santarém, three days upriver from Belem, there is a small museum in the old city hall and fine beaches at nearby Alter do Chao.

Manaus, capital of Brazil's vast Amazonas State, is best known as the former hub of the rubber empire. Sights not to miss include the opera house (Teatro Amazonas), the Jesuit-built cathedral, the municipal market and the decaying mansions of the rubber barons.

From Tefé, a town 40 hours upriver from Manaus, the Amazon now known as the Solimões continues on to Iquitos in Peru, from where it is a further 1,700 miles (2,720km) to the source high up in the Andes.

BELOW, *buying and selling the day's catch on the waterfront at Santarém*

Somewhere upriver, a giant fish leapt up out of the brownish waters, then disappeared from sight beneath the surface. A flock of birds wheeled overhead jabbering with the shrill screech of a locomotive. The early morning light danced on the distant banks, lined with interminable green.

We had left the port of Belém the previous night on a riverboat christened the *Clivia*. Painted blue and white, it was laden with 200 passengers, several tons of the local grain known as *manioc* and a giant music system blaring out Simon and Garfunkel hits. Hammocks made of coloured cloth hung on either side of the decks, lending the boat the air of a brilliant fairground. Below in the toilets, which doubled up as showers, giant cockroaches congregated with unconcealed audacity.

Of course, travelling the Amazon (Amazonas) is not like taking a river trip on most other rivers. First of all there is the searing heat and humidity. Then there are the insects – up to 10,000 of them – as well as jumping spiders, fat and hairy tarantulas and half-starved piranhas. If you survive all that, you will be attacked by some of the most potent malaria-carrying mosquitoes in the world, which cluster around the boat at night as if drawn by the promise of a Sunday roast. It is little wonder that the great 19th-century

RIGHT, *the port at Manaus, capital of the Amazon, on the Rio Negro*

INSET RIGHT, *typical sleeping quarters for many travelling on the Amazon's riverboats*

naturalist Alfred Russel Wallace, who travelled along the Rio Negro in 1851 in a dugout canoe, was driven to such distraction that bitter crystalline substances became his saving grace: 'I began taking doses of quinine and drinking plentifully cream of tartar water, though I was so weak and apathetic that at times I could hardly muster resolution to move myself to prepare them', he wrote.

My journey, I suspected, was to be of a less onerous nature. Firstly I had a cabin, situated above the rows of swinging hammocks, then I had a temperamental, but adequate, fan. Finally, if all else failed, I had a return air ticket to London dated just three weeks hence.

Leaving the Delta

After leaving the old port at Belém we sailed west through the Amazon Basin, a gigantic delta made up of countless channels, each one bigger than the next. Working our way around Marajó, an island the size of Switzerland, we passed narrow strips of forest and swampy marshlands. Soon channels like the tentacles of an octopus opened up, vast streams that tugged us gently towards the heart of the Amazon.

Many legends surround this river, including the story that its name derives from a race of strong female warriors known as the Amazons who were believed by the Greeks to live near the Black Sea. The women are said to have cut off their right breast so they could use a bow and arrow more easily. Today the world's greatest waterway bears other legends. Locals claim that its twisting waters form the great serpent mother and villagers believe that at certain times of year the dolphins, or *boto*, turn into men in white suits and are responsible for pregnancies before marriage.

It was late afternoon and the light was softening when we reached the village of Gurupá.

BELOW, *the Opera House in Manaus, part of the rebuilding programme carried out in the late 19th century as a result of the rubber boom*

SILVA N

ABOVE, *the sublime to the ridiculous ... there is no limit to the size of craft on the Amazon – at either end of the scale*

From a cluster of palm-thatched huts on the river bank, a flotilla of canoes sped towards us. Propelled with energy bred from desperation, the paddlers waited in mid-river and as our vessel passed by they sprang forward, gripping the moorings of the boat with their bare hands. The diminutive Indians then clung on whilst passengers threw food or bundles of clothes down to them, finally letting go and drifting back downriver until they were little more than dots on a distant horizon.

Downstairs in the galley, next to the engine room, dinner comprised of beans, rice, spaghetti and fried legs of chicken was being served. The air stank of diesel fumes and the table vibrated gently to the pounding of the motor. We held our breath and piled our plates up high before escaping back on deck, back where the air was fresh and the *Clivia* hugged the outline of the shore.

A Mighty Waterway

To imagine the Amazon is to conjure up an immense new world. In total, this mighty river extends over an area equivalent to 7 per cent of the earth's surface. At times, it stretches out on all sides; it disappears into distant horizons; it literally swallows up the land. Around 1,100 tributaries feed into the Amazon, including the giant Rio Negro, the Rio Japurá, the Rio Purus and the Rio Madeira, and together they account for one-fifth of all river water.

Sailors claim that westbound ships still 75 miles (120km) out to sea run into the Amazon's vast muddy waters.

The mighty Amazon is a river of subtle moods, dictated by the seasons. Every year, when the torrential rains come in December, the river bursts its banks and water spreads over thousands of square miles of flood plain. These plains support countless species of plants and animals and it is estimated that the Amazon contains more than 1,800 species of butterfly alone, as well as swimming snakes and fish with two sets of eyes.

Awed by the scale, exhausted by the humidity, chased by mosquitoes, there is little with which to compare it. Only the steady chug of the boat breaks into my reveries; it punctuates the minutes which turn into hours, and the hours which turn into days.

The first Europeans discovered the Amazon more than four centuries ago. On 11 February 1542, the Spanish conquistador Francisco

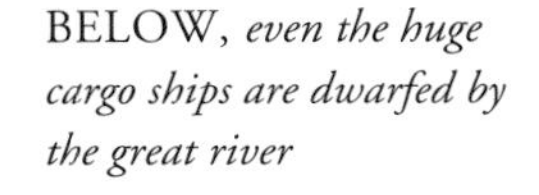

BELOW, *even the huge cargo ships are dwarfed by the great river*

de Orellana and his contingent of 60 men rowed out of the Napo River into the vast reaches of the Rio Negro. Sailing downstream, they crossed the Japurá River then continued towards the Amazon Estuary, reaching Marajó Island in mid-July.

Friar Gaspar Carvajal, who accompanied the expedition, kept a detailed diary of his journey. 'The Amazons go about naked,' he wrote 'but with their privy parts covered, with their bows and arrows in their hands, doing as much fighting as ten Indian men.'

For centuries the Amazon basin remained shrouded in all sorts of make-believe. It was a place thought to be inhabited by headless people, and some men even claimed to have come across a race who had their feet turned backwards, so that pursuers would track them in the wrong direction.

Eventually, however, the region took on a different character. First the Spanish and later the Portuguese and Brazilians plundered the forest; vast areas of land were cleared for rubber and for sugar plantations. The pitiful story of merciless development and exploitation continues even today. Each year an area the size of Great Britain is logged to make way for cattle ranches which are largely funded by government subsidies. In parts of Amazonia, mines have sprung up to extract gold, uranium and other mineral resources. The mines have brought undreamt of wealth for the few big landowners, but they also cause untold pollution. Every year, 13,000 tons of mercury from gold mining finds its way into the Amazon River. The future is not a bright one: at this speed, without large-scale reforestation, the last tree in the Amazon forest will be cut down little more than 60 years from now.

DAY AFTER DAY ...

The light of dawn brings the first awakenings from the torpor. We had sailed through the night past a dull outline of land, the smell of cattle manure mixing with the fresh river breezes. Now the river has widened. We stare out on to grazing pastures and small Indian villages with neat white churches.

Down in the galley, my fellow passengers are queuing for breakfast. They stretch their legs painfully after hours in their hammocks and clutch at diminutive children, some little more than a few months old. They literally fight for biscuits spread with acrid butter and for cups of sweetened coffee drunk out of plastic cups – then discarded into the river.

Upstairs, the few Europeans have had a no less eventful journey. An American named Mark was awoken by a bat which flew into his

BELOW, *local people waiting for the boats to arrive at Alimeirim quay*

FAR RIGHT, *a typical Indian village on the river bank; the water is essential for washing and cooking*

INSET, *the banana-laden canoe of a lone trader on his way to market*

cabin. It got caught up in the fan and was shredded all over the floor. Others complain of blocked-up toilets or the sound of techno music blaring out over the ship's speakers.

At Santarém, a busy little port at the junction of the Tapajós River, we spend the day sightseeing and eating delicious Tambaqui fish, freshly grilled on red-hot coals. In the evening, a Brazilian anarchist, who sleeps rough in the square, joins us at our table near the port. He decries the corrupt government, the lack of support for the poor people of Brazil and the greed of the rich landowners. Finishing his beer, he asks for money. Then, empty handed, he wanders unsteadily off into the night.

We escape aboard our new vessel, the *Moreira Da Silva III*. It has a car on the bottom deck and half a cow hanging outside the bathroom – our food for the next leg of the journey. In every corner of the vessel passengers in hammocks are stacked up like tins of cut-price tuna on a supermarket shelf.

From Santarém we continue upstream, up past the two-coloured confluence of the Tapajós River and towards the town of Óbidos. Once again the monotony takes over, measured only by the draining heat and the number of squashed cockroaches.

One night, an anaconda snake is found on board our vessel. About 6ft (2m) long with beautiful colourful patterns, it slithered along the deck under the hammocks and disappeared into the hold to the sound of frantic screaming. Only later did we learn that it belonged to the captain and was kept along with two other snakes to keep the rats in check.

On another night, a drunk claiming to be a federal policeman came on board. He demanded our passports and said that all foreigners must leave the country, at which point we retreated into our cabins and locked the door. After knocking for five minutes he eventually lost interest, descending at the next port where he was swallowed up in the darkness.

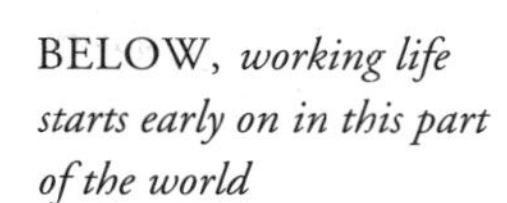

BELOW, *working life starts early on in this part of the world*

Hugging the bank to avoid the full force of the current we continue slowly upstream, passing the towns of Oriximiná and Juruti. Now and again we come across motorised canoes or giant cargo boats carrying trucks downriver. Once we drew alongside a vast oil tanker on its return voyage from Manaus. All throughout, the landscape rarely changes, the soft grassy meadows and sparse treeline broken only by small towns and villages, short stops on our seemingly endless journey.

JUNGLE BOOM TOWN

On the sixth night we see lights in the distance. The river traffic increases and ahead there are concrete houses and the distant outline of high-rises seems to sprout up out of the jungle. After almost a week on the river, we have reached our first major centre of civilisation.

There is an opera house in Manaus. It has Italian porches, baroque-style colonnades and a dome covered in green-blue and gold tiles. Opened in 1896, it is the most bizarre symbol of the unprecedented economic boom that transformed this isolated riverine city into one of the world's wealthiest metropolises. During the mid-19th and early 20th century, demand for rubber to make motor car tyres sent prices spiralling in Amazonia. Up to 80,000 tons of rubber a year were shipped from Manaus to ports as far afield as New York and Liverpool and overnight this town at the confluence of the Amazon and Rio Negro grew into a city of 50,000 inhabitants. Electric trams were built, so too the opera house where ballet troupes came from Europe to perform to diamond-bedecked audiences. At one stage it was even fashionable for rubber magnates to send their laundry to Europe to be cleaned.

But the wily British smuggled out seeds which were planted elsewhere in South-east Asia and soon rival rubber plantations sprang up in Malaysia and in Ceylon. In 1912, as suddenly as it had begun, the rubber boom ended. Landowners committed suicide, the electricity was closed off and the opera house temporarily closed down. Now, however, a new wave of prosperity has gripped parts of Manaus. Since the government declared the area a free-trade zone back in 1967, multi-nationals using cheap labour have created a new source of income. In the midst of the world's largest rain forest, more than 40,000 workers make television sets, watches and motorbikes which are distributed throughout Brazil and Latin America. But the incongruity is striking. In the shadow of high-rises are some of the worst shanty towns in Brazil. It is a frightening picture; a picture of the world gone badly wrong.

ALL CHANGE

In the busy riverside docks at Manaus, where the waters can rise by as much as 90ft (27m) during the annual floods, we change ship. Our

ABOVE, *the market at Tefé, a busy town in the heart of the jungle*

new home is a crowded three-decker vessel called the *Fernandes*. It has brown and white awnings, a satellite dish on the owner's cabin – and a bar piled high with Antarctica beer and toasted sandwiches. In the hold below, passengers busily load up timber, bags of cement and a chicken on a lead. The crowded dirty decks are filled with traders, workers looking for a new start in life, a pregnant girl fleeing her parents. The cast is different but the faces are the same: faces that show resilience and strength, the strength to continue against the hardships of life on the river.

Leaving Manaus, the *Fernandes* manoeuvres into the main stream and the distant *favelas* (shacks, shanty towns) give way to factories and then to farmland. Soon the river changes, the black waters of the Rio Negro merging with the brown waters of the Amazon, at this point measuring up to 6 miles (10km) wide and now known as the Solimões.

Slaves were once brought to the confluence of these two great rivers from as far afield as Africa to work the sugar plantations. Many died on the journey, but others survived to intermarry with the local people, the Indians and the Portuguese, producing that rich mix of ethnic races that is the true hallmark of the Brazilian people.

Here and there, sandbanks rise out of the murky waters of the river, gargantuan islands which for six months of the year are cultivated by the local people and for six months are flooded. If experiments carried out by the Instituto Nacional de Pesquisas da Amazonia (INPA) are anything to go by, these fertile soils could provide sustenance for villagers not only in the Amazon, but in other parts of Brazil.

At Codajás, when the boat draws alongside the quay, a handful of passengers leap ashore clutching hammocks and cheap polyester bags.

Further on the rain forest once again muscles in on all sides. Occasionally we glimpse flocks of birds or parrots swooping down from the lush vegetation along the river banks. Once we even saw dolphins as they leapt, their beautiful arch-like movements breaking the surface of the slow-moving water.

THE MARCH OF PROGRESS

Nights on the Amazon are beautiful and tranquil. A fresh breeze blows and the distant shores are silhouetted by the light of the moon. Dusk is also a time of reflection. More than 2 million Indians once lived along these river banks in harmony with their environment and their story is a tragic one. Murdered by the conquistadors, ravaged by smallpox, tuberculosis, flu and measles and driven from their land by ranchers and politically connected businessmen, they have become fodder for the giant animal known as progress. Today, fewer than 200,000 Indians live in the Amazon and many of these are dependent on handouts from illegal gold miners. Others are being driven even further from their homelands by new roads, by the relentless advance of logging companies and by pollution of their rivers.

By the time our boat leaves the port of Coari, the sun is already arching behind the horizon. Once again we contemplate the monotony of the land, a land that has been ravaged by man in the name of progress.

Soon, however, the clearings dissipate. Along the river banks the trees grow to heights of more than 100ft (30m), their trunks occasionally as large as houses. The further we go, the fewer villages we see and the jungle envelopes us.

On the last night a storm blows up. The sky turns black and a tropical downpour ensues. Canoes and primitive wooden dugouts scurry along the water in search of shelter, while a few Indians squat outside their houses, staring out over the angry waters.

Nearing the end of my journey, the mood changes. Passengers smile after their long days in the hammocks, the television blares out news of Brazil's latest football conquests. Here too the river takes on a different guise, a certain wildness and untamed grandeur as it continues on its long route through Iquitos to its source high up in the Peruvian Andes.

On the last morning the sun is gentle, the river banks lush and full of promise. A few dolphins leap out of the water. The boat manoeuvres against the pier at Tefé, a lively riverine town with a raw frontier feel about it, the feel of being in the midst of the Amazon. It is a fitting end to my 10-day voyage, a journey that will always carry poignant memories of the power and fertility of the river – and the destructive forces of civilisation ranged against it.

PRACTICAL INFORMATION

- Boats between Belém and Manaus depart daily. Generally, the 1,000-mile (1,600km) journey takes about five or six days. From Manaus another boat runs to Tefé, 400 miles (640km) further upstream, taking about 40 hours. It is also possible to continue to Iquitos in Peru, although you will need a visa and the constitution of an ox.
- The best time to travel is between July and November, when rains average just 20in (50mm) a month. From December to May rains can be as high as 118in (300mm) per month. At all times the climate remains extremely humid.
- Most boats are overcrowded, ancient and uncomfortable. Accommodation is in hammocks or cabins. Meals are included.
- Visitors to Brazil from EC countries do not require visas. You will, however, need injections including yellow fever and will require anti-malaria tablets. Ask your doctor for details in good time before leaving.
- For information about trips contact Last Frontiers, Swan House, Long Crendon, Bucks, HP18 9AF, Tel: 01844 208405; Fax: 01844 208405, or Trailfinders, 194 Kensington High Street, London W8 7RG; Tel: 0171 938 3939; Fax: 0171 938 3305. In Australia Trailfinders, 91 Elizabeth Street, Brisbane 4000, Queensland, Tel: 07 3229 0887.

LEFT, *the port at Tefé, journey's end*

The Great White South

ANN F STONEHOUSE

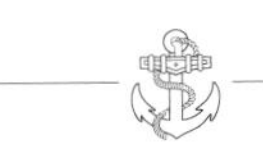

ABOVE, *some 40 species of bird breed in the Antarctic, including the black-browed albatross*

Antarctica – a land of ice and heroes, peculiar wildlife and a growing ozone hole. As the daughter of a polar and penguin expert I'd somehow lived with this icebound continent all my life, but never believed I would get there and see it for myself. The opportunity arose to fill in as a research assistant for three weeks, and suddenly it was all real and happening. The prospect of seeing this land with my father was thrilling but also terrifyingly outside my experience – and how would I, who can feel sea-sick in a row-boat, cope with three weeks on the wild Southern Ocean?

Delayed briefly by a gale, we sailed from Port Stanley on a glorious calm, sunny morning in mid-December. The Falklands had reminded me of some of the more remote parts of Scotland, but as I looked back from the ship's quarter-deck I saw a clutch of black and white Magellanic penguins in a sandy bay. The familiarity was a lie, and I really was heading for new territory.

We were travelling on the graceful little Russian tour ship MS *Alla Tarasova.* Her 320ft (100m) steel hull was supposedly ice-strengthened, and I wasn't the only passenger to glance at it anxiously as we climbed aboard, looking for the reinforcement. Under her recent coat of paint, she was giving nothing away.

BELOW, *a crane and winch on the* Tarasova's *bows allow for speedy transfer of the Zodiacs*

We had two days at sea to get used to the ship, its informal routine and our 60 fellow passengers and cruise staff. These included a geologist and a naturalist who gave lectures, and an expert ornithologist, on hand to identify the ever-present seabirds.

At first I thought the birds were following the ship – then I realised that they lived over the water. We were simply travellers passing through. They swooped and hung low over the quarter-deck, at ease and curious, controlling movement with the slightest wing tilt – giant petrels (stinkers) with their friendly, ugly faces and hooked pink bills, beautiful black-browed albatrosses (mollymawks), wandering albatrosses with their vast, slender wing span, and cute chocolate-headed pintados.

Lessons in basic Russian language, offered by the purser, helped to pass the time at sea and aid communications with the friendly crew, far from their home port of Murmansk. We were welcomed officially at a 'Cocktail Party'. The captain himself, splendid in gold braided uniform, bowed low over my hand as the ship rolled and we all struggled to hold our feet. 'Beautiful!' he murmured, with one eye on my

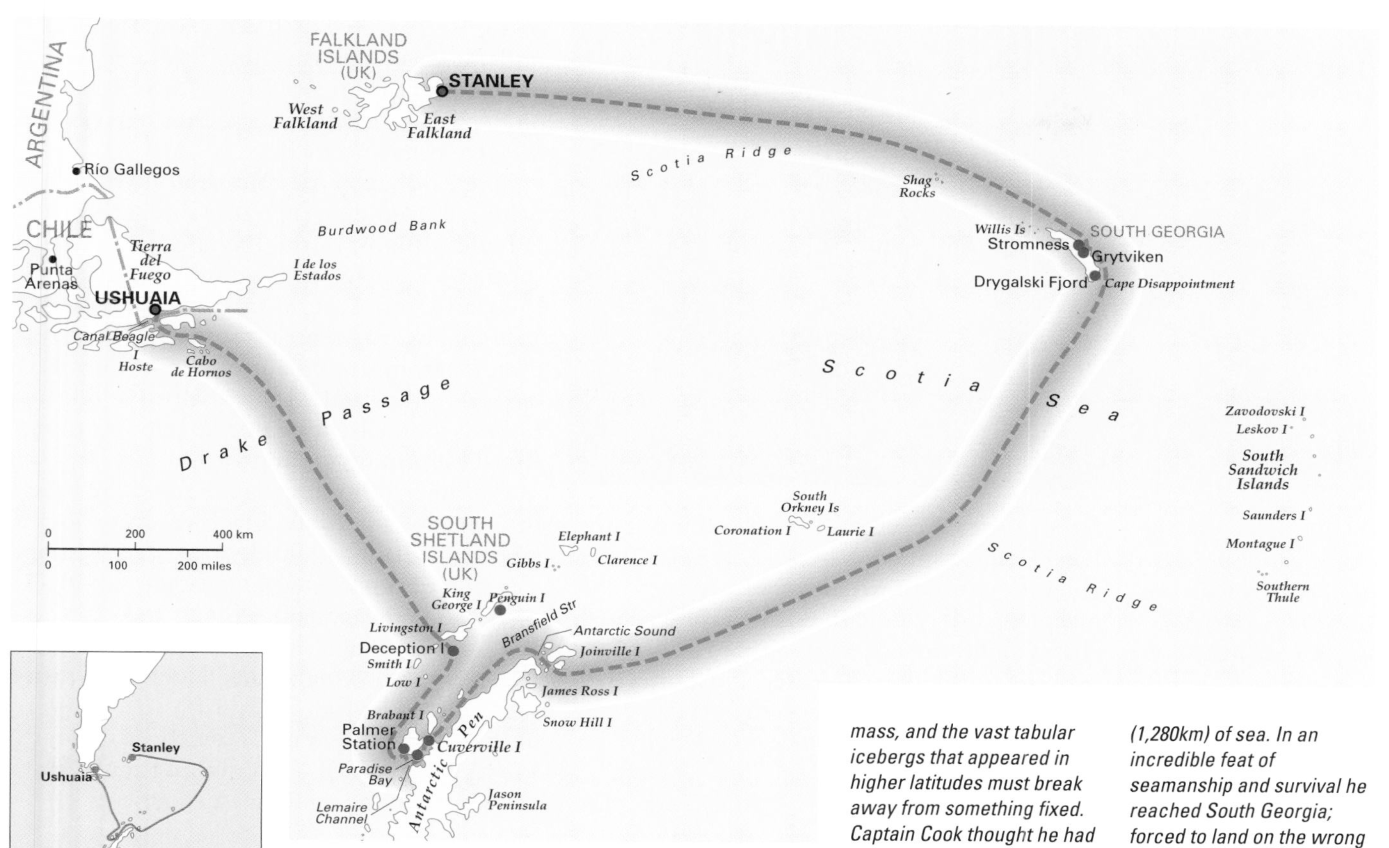

The Antarctic Peninsula is the long finger of land stretching up towards South America. Despite appearances on the map, it is not a geological extension of the Andes, but swung off the end of South Africa millions of years ago. Many countries have claimed territory in Antarctica in recent years. The Antarctic Treaty, first set up some 35 years ago, provides a framework for international scientific co-operation on the continent, while avoiding issues of sovereignty. Antarctica is a growing tourist destination, with some 8,000 visitors a year passing through. Most tours are run safely and observe ecological guidelines, but at present there are no regulations to ensure this.

I quickly learned that my own voyage of discovery was haunted by the ghosts of those who have gone before. The earliest explorers guessed there was land at the south pole – the cold air, they argued, must sweep off a continental land mass, and the vast tabular icebergs that appeared in higher latitudes must break away from something fixed. Captain Cook thought he had discovered the continent in 1775, but rounded the tip of South Georgia to find that this was but another island. He named the point Cape Disappointment and sailed away, driven off by encroaching ice and storms. Shackleton lost his ship in the pack-ice in 1916. His crew reached Elephant Island, and he set off in an open boat in search of rescue, across 800 miles (1,280km) of sea. In an incredible feat of seamanship and survival he reached South Georgia; forced to land on the wrong side of the island, he had to climb the central mountainous ridge to reach Stromness and safety. Rescue expeditions were launched from South America, and remarkably not one man was lost.

BELOW, *off-loading eager explorers in the Bay of Isles*

ABOVE, *Zodiacs allowed passengers to get close to the water, which in this sunny channel on South Georgia supported tiny, transparent comb jellyfish*

father to see the effect. I simpered modestly. Later I caught him doing exactly the same to an elderly matron in white cowboy fringes – his one sure word of English was well chosen! Such charm and formality was forgotten in the daily running of the ship, however, much of which we watched at first hand on the bridge, easily the best place on board for clear viewing.

Sight of Land

Indeed, it was from there that I got my first view of South Georgia. The extraordinarily clear air and bright light meant that we saw the mountainous island from many miles away, minus its usual blanket of cloud. After the open sea it was a relief to see land – even these inhospitable peaks of brown rock, mottled with snow – and I thought of Ernest Shackleton making his way here in an open boat in 1916.

Tentatively at first, we took to our own ships' boats – black rubber Zodiacs – and cruised around the Welcome Islets, high black rocks crusted with colourful pink and gold lichen. Somehow a colony of macaroni penguins lived on top and we eyed each other, we from under unflattering woolly hats, they from under their beetling yellow brows. Young fur seals basked on the ledges at sea level, and they too watched us, largely unconcerned.

BELOW, *the king penguins nesting at Salisbury Plain number tens of thousands*

The Zodiacs came into their own as they delivered us straight on to the shingle beach at Salisbury Plain, in the Bay of Isles. We had spotted king penguins swimming in the shallows and standing along the shore as we approached, and were astonished to discover that, far from fleeing our presence, they were curious and unafraid, strolling solemnly towards us for all the world like some morning-suited welcoming committee.

On the flat, muddy plain behind the beach little groups and individual birds were dotted around, many in an untidy state of half-moult. Temporarily unable to go to sea, they stood disconsolately in stagnant puddles to keep cool. We passed through slowly, anxious not to disturb them, following a track flattened by many small feet. The first indication that we were nearing the colony was the stench – a curiously strong, meaty smell – but nothing could really prepare us for the sight that met our eyes.

Penguins, Penguins, Everywhere

The colony was set well back from the shore, a vast, noisy, tightly packed mass of penguins in charcoal, grey and gold, which stretched far up the distant hillside. I sat down quietly on a rock to take it all in.

Now and again the wind would whip up a miniature dust storm of dried mud and curly, shed feathers, and we all – penguins and observers alike – shut our eyes and choked. Everyone ducked, too, when a wicked-looking skua flew in low overhead. Little white birds pecked among the penguins, looking like doves but with peeling, raw faces and vicious bills. These were the sheathbills, scavengers who, like the skuas, live on the extensive detritus of a colony this size.

There were penguins preening, parading up and down to the sea, dozing upright while leaning back on their bristly tails, or fast asleep on their stomachs with their flippers neatly tucked underneath. A few sat with a tell-tale bulge, and my first glimpse of a parent bird turning the great white egg on its feet was a tremendously exciting moment.

The other species we saw in large numbers on South Georgia was the much smaller, shier chinstrap penguin, distinguished by its little black helmet and red eyes.

On arrival at Cooper Bay my task was simply to count the number of nesting pairs, and I set

ABOVE, *seals of the southern seas were once hunted almost to extinction, and the population is only now recovering*

RIGHT, *in summer the sea-ice breaks up to allow access to the spectacular Lemaire Channel*

BELOW, *the density of the ice gives this berg, in Cuverville Bay, its extraordinary blue colour*

ABOVE, *Grytviken was once a bustling community, dedicated to processing whale meat, bone and blubber for its rich oil*

BELOW, *king penguins lay their eggs between November and March, and the chicks are dependent for over a year*

off confidently through the tall tussock. I had to keep low, so as not to frighten the birds, but had also to avoid stepping on the elephant seals who snored there. I nearly lost a boot in a particularly smelly seal wallow – this was harder going than I thought – and then suddenly plunged up to my waist in a stinking, fetid pool. I had caught hold of the grass as I went down – just as well, because my feet still weren't touching the bottom. Perhaps it had no bottom. Horrified, I hauled myself out on to the tussock, thankful I hadn't fallen headfirst. I decided not to risk a real accident and stood up where I was, counting penguins. My eyes smarted in the cold wind, and after a few false starts I had reached 800 when my father found me. We doubled that to estimate the complete size of the colony, before returning to the ship and some ribbing about my unpleasant odour.

One Moonlit Night

I got another fright in the empty wooden buildings of the old whaling station at Stromness. We had anchored in the bay on a beautiful still, moonlit night, and people swore they saw ghostly lights moving in the deserted buildings. It was probably our own lights reflected in the broken windows, but as I stood alone just inside the old machinery workshop the next morning and heard a deep, shuddering sigh, a chill ran down my spine. I held my breath, but not so my 'ghost', who sighed again. I took a step further in, and almost fell over a great blubberous elephant seal, sublimely asleep amid the wreckage.

The seal populations have gradually been restored after the severe hunting that took place here from the 18th century, but the whales have yet to recover. The much bigger processing plant along the coast at Grytviken was still operating up to 1965 and my father recalled seeing it in full swing, when the air was thick with the stench of cooking whale. A whole community lived at Grytviken, and although the old cinema has finally blown down, the wooden Norwegian church is restored and delightfully complete with wheezy harmonium.

There is an excellent little museum, too, in the former governor's house, showing the way of life for the men who lived and worked here. Polar hero Ernest Shackleton is buried in the graveyard, but the more dramatic memorial to him stands on the other side of the bay, on a site entrenched and mined by the Argentines during the Falklands War and now surrounded by the British garrison. We welcomed a party of soldiers on board one evening, along with the young harbourmaster and his wife, who loved the life on this remote island at the bottom of the world.

Christmas in Paradise

After a sunny morning visit to the tumbled ice of the Drygalski Fjord, we left South Georgia behind and headed for the Antarctic Peninsula, celebrating in unlikely fashion with a shipboard barbecue and a thigh-slapping dance from our Austrian chef. The seas were bigger now, and the foamy green water swirling over the porthole of my little cabin made it feel like the inside of a washing machine.

Two days later, as the ridges of the South Orkney islands came into view, we saw our first proper icebergs. They floated by in all shapes from needle points to flat tables, increasingly bigger, and sometimes streaked with the incredible, vivid turquoise of ancient ice. As we travelled on southwards the weather became much colder and windier. On a sunny night we sailed down the Antarctic Sound, a gale blowing ice-crystals outside, and the low light glowing on the magnificent peaks of the Peninsula.

Christmas Eve, however, dawned warm and sunny, with not a breath of wind – the meteorologists had never recorded a day like it. We had a warm walk across the volcanic cone of Penguin Island, with its ancient, tufted lichens and rocks crazed by frost and polished smooth by the wind. Antarctic terns screamed overhead as we scrunched carefully past their nesting sites, and a lone Adélie tobogganed unhurriedly down a slipway of old snow, keeping pace with us. I began to understand why my father is drawn back to this beautiful land year after year.

Our expedition leader had promised us Christmas in Paradise, and we still had some miles to travel. We were underway at 5am on a bright Christmas morning, heading along the Bransfield Straight. Just a few of us were on the bridge to see the best present of the day – three minke whales diving high out of the water to feed. By breakfast time we were approaching Cuverville Island, where my father's team of PhD students was camped. Laden with mail from home and gifts of festive food from the galley, we were sure of a warm welcome. The five-year programme here, monitoring the effects of tourism, centred around an extensive gentoo colony; visitors were requested to leave half the birds alone, as a control, but were welcome to visit the other half, within the usual careful guidelines. Stress on the penguins was measured using a specially developed electronic egg, and so far the results were surprising – the penguins visited by tourists were doing rather better than their isolated neighbours.

I was reluctant to leave Cuverville but Paradise was calling. Paradise Bay is dominated by a deserted Chilean survey station, built on the historic site of Waterboat Point. This was named for two hardy British lads who in 1921 stayed on in an old wooden boat to complete the job after their companions had left for the winter months.

We had one more sunny day in Antarctica before our luck ran out, and that was spent cruising in the magnificent Lemaire Channel. We went on to visit to the American Palmer Station (which offered our first and last Antarctic souvenir gift shop) and turned resolutely north, passing through the Neumayer Channel in pink, soft evening sunlight.

By the time we reached the volcanic crater of Deception Island the wind outside was much too strong for us to land, and we only just made it through the narrow entrance known as Neptune's Bellows. As the wind howled down on us, I took my fiddle out on to the sheltered, still sunny side of the deck and played screaming Shetland reels into the gale. Within a few minutes I and the instrument were covered in gritty black lava dust. Despite its sheltered position, the ship dragged its anchor that night and fouled on an ancient chain: next morning a diver cut us free, and between the gusts we made our getaway, heading for home.

Now the sea was really rough, and as the ship lurched and rolled your bunk was the safest (and in my case the only) place to be. I prefer to draw a veil over the next 48 hours as we sailed across the heaving Drake Passage, the roughest stretch of water in the world, where the icy polar waters meet the warmer northern seas. On the last morning I awoke as from a dream at 3am. Outside it was dark, which after the midnight sun felt very strange. The ship had stopped rolling – we were in the Beagle Channel. By 5.30 I was ready to make up for lost breakfasts, and enjoy the sight of the green slopes of Tierra del Fuego slipping past in the morning mist. We were heading for the red roofs of Ushuaia, lit up in its own brilliant ray of sunshine.

LEFT, *Ushuaia, once a penal colony at the end of the earth, is now the busy gateway to Antarctica*

PRACTICAL INFORMATION

■ The Antarctic summer season runs from November through to March. Weather can vary enormously, but is generally cold, dry and windy, with temperatures below freezing – we were particularly lucky with sunshine and calm days.

■ Take your passport, binoculars and a camera. The right clothing is vital, and removable layers are recommended. Tour operators will issue advice, but a warm, windproof Parka is essential, as are waterproof trousers, a warm hat and mittens, and rubber boots with thermal insoles for going ashore. Take sunglasses, as the glare can be intense.

■ A polar visit is expensive, and for maximum enjoyment it's worth reading up before you go. Recommended: Shackleton's historic account, South; Bernard Stonehouse's general introduction, *North Pole South Pole* (1990); Tony Soper's guide to *Antarctic Wildlife* (1994).

■ Companies around the world offer tours to different parts of Antarctica, lasting from six days to three weeks. Choose your ship carefully – ours was ideal, big enough to get through the rough seas in some comfort, but small enough to manouevre into tight bays, for everybody to land quickly, and for a friendly atmosphere on board.

■ I travelled with Quark Expeditions, 980 Post Road, Darien, CT 06820, USA. Tel: 203 656 0499. Fax: 203 655 6623.

Cruising the Nile in the Wake of the Gods

ANTHONY SATTIN

ABOVE, *taking a breather before the next batch of tourists arrive for a ride*

Ancient Egyptians called their land Kemet, which Arab writers corrupted to al-Kam and described as a place where people turned ordinary metal into gold. From *al-Kam*, we have the word alchemy and from this journey along the Nile through Egypt, we still get magic. The *Nile Serenade* is part of the Thomas Cook fleet, the company which pioneered Nile tourism in the 1860s. From its decks I saw the contrasts for which Egypt is famous: hard-won farmland and encroaching desert, brilliant days and star-lit nights, mud-brick villages and, most astonishing of all, some of the world's most spectacular ancient ruins.

A 19th-century Frenchman by the name of Ampère, son of the man who gave his name to the unit of electricity, described a journey in Egypt as being 'a donkey-ride and a boating-trip interspersed with ruins'. Soon after we arrived in Luxor – at night by plane from Cairo – we boarded our cruiser, the *Nile Serenade*, moored a little to the south of Luxor town.

The *Nile Serenade* is one of the new Cook's fleet. Launched in 1990, its 21 cabins and six suites are more comfortable than anything that Ampère could have dreamed of in the 19th century. I slept well and woke to find the Nile outside my window in all its morning-silver, hyacinth-choked glory. Across the river, the bank was lined with palm trees and the Nile's valley was

BELOW, *a typical Nile cruiser at Luxor*

The Nile valley is broad at Luxor, but across the green line, the mountains of the Eastern and Libyan (Western) deserts can be seen. Heading south, these mountains and the desert they hold back close in. At Aswan, the green strip is no more than a slender field, wedged between desert and river.

The greatest spectacles along the river are the remains of the ancient temples. Along the river, temples on both banks at Luxor, at Kôm Ombo and Aswan can been seen from the boat and all of them, as well as the temples at Esna (Isna) and Edfu (Idfu), are worth waking up for.

For as long as people have lived along the Nile, the river has been a means of transport and communication. The ancient Egyptians travelled by boat (one of the finest of these is preserved in a special museum beside the Pyramids of Giza), sailing upstream with the prevailing winds and floating down on the current. In the 19th century, British engineers extended the railway line from Cairo to Aswan to speed the movement of troops to the relief of Khartoum in the Sudan and a road now runs alongside the line. Although most goods are now carried by land, occasional barges laden with stone, gravel or fruit are still seen on the Nile.

There are two quite separate dams across the Nile at Aswan. The first was opened in 1902 and, at 1 mile (2km) across, was the largest in the world at the time. The second, built upstream, was opened in 1971. More than 2 miles (3.5km) long and standing 365ft (111m) above the river bed, it has created a lake which stretches 312 miles (500km) back into Sudan. The lake ensures a year-round supply of water to irrigate the fields (rain is rare in most of Egypt). For tourists, it reduces the risk of boats running aground on shallows. It also appears to be changing the climate, bringing unusual rains to southern Egypt.

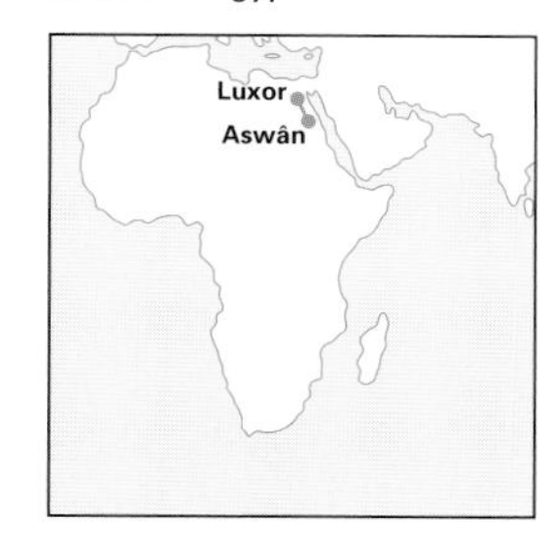

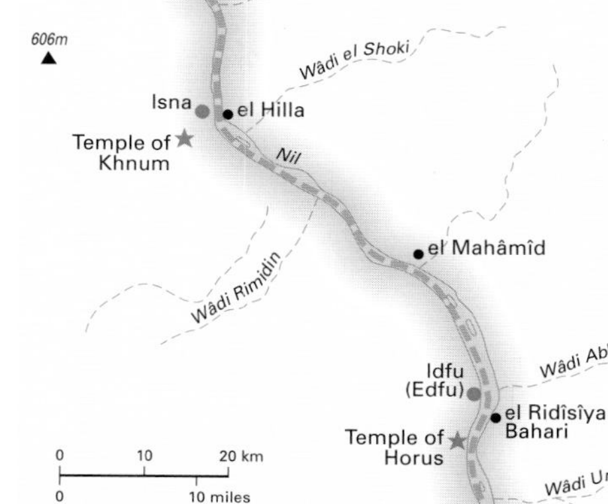

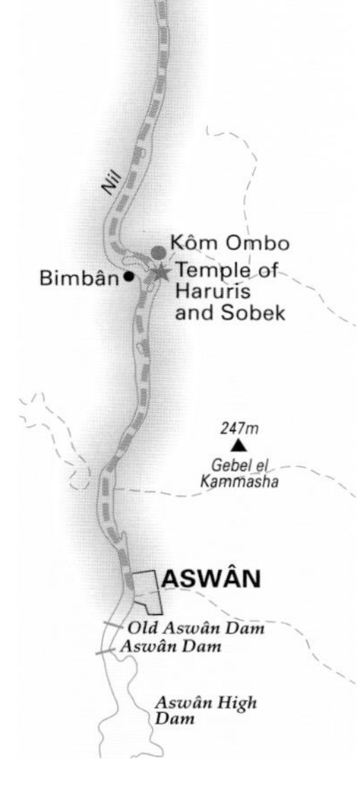

BELOW, *tempting herbs and spices can be seen on display in every Egyptian bazaar*

carpeted with as brilliant a green as nature can make. Beyond the fields, the great Theban Hills, the graveyard of pharaohs, caught the sun. It was the sort of view I could have stared at – and lost myself in – all morning, but on a Nile cruise there is a strict timetable to be kept and breakfast was already being served.

THE WEST BANK

The Nile is a great unifier. Many thousands of years ago the world went through a 'global warming' and the forests of north-east Africa were turned to desert. The inhabitants of the region, who had until then lived off whatever they could hunt or collect, were forced to adapt to survive. Without rain to provide them with food, they relied on the annual flooding of the Nile to irrigate the land. Congregated along the banks of the river, they learned to live with each other and in the process developed what we now call civilisation.

The Nile had a similarly unifying effect on passengers aboard the *Nile Serenade*. Mostly British, the previous evening we had been strangers in a strange land, but now we were part of a group, passing milk or butter, sighing at the one amongst us who most liked to complain – about having a cabin with a land-side view, a wardrobe without hangers, a headache from the beer – and straining to catch a glimpse of the couple who had signed up for the hot-air balloon ride. Lost in the morning mist, perhaps even now they were drifting over the Valley of the Kings and trying to keep down their early-morning champagne breakfasts. Before the sun was much higher in the sky, we were ferried across to the West Bank to join them and it was there that we found the second of Ampère's elements, the ruins.

Ancient Egyptians believed that the dead lived in a parallel world situated somewhere beyond the Western Desert, where the sun disappeared at the end of the day. For this reason they were invariably buried on the west side of the Nile. The funerary cults of the ancients are too complicated to grasp in a single morning spent rushing in and out of tombs and temples, but this basic idea – of a geographical progression from east to west, from the promise of the rising sun to the shadows at

ABOVE, *the Valley of the Kings, with its amazing underground world, on the West Bank at Luxor*

BELOW, *Horus' boat at the temple of Edfu, which is dedicated to the falcon god*

dusk – seemed perfectly sensible to us. Climbing down into the tomb of a long-forgotten pharaoh – with the Complainer wishing that he could find someone to do as good a paint job back home – we followed the dead man past trials, traps and demons until he came to the underworld where Osiris, the presiding god, was waiting to weigh his heart against the single feather of truth on his scales of justice.

A morning spent in this environment had an unexpected influence on us. Our journey that day had been more through time – back into the world of the ancients – than across the land (we hadn't travelled very far), but when our guide Norma referred to a rock above the Valley of the Queens as 'the goddess's womb' and talked about 'the breaking of waters at the primeval moment', even the most down-to-earth amongst us seemed to understand.

'It's easy to think of the ancient Egyptians as being obsessed with death', Norma said when we were back on the boat having lunch, watching the crew prepare for our departure. 'Most of the buildings that survive are to do with religion and death. But all most ancient Egyptians wanted to do was to live well.' She looked at the waiters serving huge quantities of food and laughed. 'Just like us now.' It was hard to disagree.

Leaving Luxor

That night, my second on board and first on the move, I was woken by a clunk, a shout and the sound of running water. I was frantically trying to remember where the life jackets were stowed when I looked out of the window and saw that we were passing through a lock: this was the barrage which blocks the Nile at Esna (Isna). I slept again

and didn't wake until around 8am. Pulling back the curtain I found a more remote land going by than we had seen at Luxor. The colour of the water and of the life alongside it was as rich as the colours of a tomb painting. The sun dazzled the tree tops, the sugar cane was high, buffaloes were brought down to the water and fishing boats were still out, two men in each, one rowing while the other cast his net. The sight of the men at work shook me from my slumber.

Of the three significant surviving temples we visited between Luxor and Aswan – at Esna, Edfu (Idfu) and Kôm Ombo – I found Esna the least impressive. Although there was certainly a temple on the site in pharaonic times, all that has survived is a hypostyle hall (the roof supported by pillars) built by the Romans, merely an antechamber to the main temple, which has unfortunately disappeared. Edfu, on the other hand, was something else entirely.

We left the *Nile Serenade* by *caleches* (old horse-drawn carriages) and rode in brilliant sunshine through Edfu town. These carriages are common in town – even Egyptians use them – so few heads turned as we passed. When we reached the temple we found it so well preserved that we were thrown back even further in time, several thousand years in fact, for this was Cleopatra country, evoking memories of the ancient times.

GREATEST OF THE TEMPLES

Edfu Temple was dedicated to Horus, the falcon god. According to ancient mythology, it was he who avenged the death of his father, Osiris, and fought his way back to his rightful place on the Egyptian throne. The story is a parable of the struggle between chaos and order and it is clear from the temple's symmetry and layout which side Horus was on.

Edfu Temple is one of the best preserved ancient monuments in Egypt. It is also one of the best documented, its inscriptions giving an unusually complete history of the building. There had been at least one earlier temple on the site, but the one we saw wasn't started until 237 BC. Its main building was completed in 212 BC but then a further seven years were taken up decorating the walls before the great doors were fitted. Even then, it wasn't finished. Because of uprisings in that part of the country, the temple wasn't fully decorated and equipped for a further 65 years; in 142 BC it was finally dedicated. Ninety years later Cleopatra ascended the throne in Alexandria and would certainly have visited Edfu during her reign.

Two thousand years after the pharaoh's artists finished work at Edfu Temple, we can still make out some of the nature of their beliefs, the love of order, the need to appease many different gods –

ABOVE, *hieroglyphics and pictorial representations adorn most temple surfaces*

MAIN PICTURE, *view of the Nile, complete with feluccas, from the Old Cataract Hotel, at Aswan*

each one of which represented a different feature of the natural world (gods of the sun and moon, the river and so on) as well as of the human characteristics (gods of justice and truth).

It was hot, our throats were dry and the Complainer was getting bored. 'Once you've seen one of these pagan temples you've seen them all.' But Norma kept up interest by pointing out things from these 'pagans' which found their way into Christianity. 'Even Horus, son of the god Osiris, is a figure with parallels in Christian teaching. He is the child sent to redeem the world, to claim his rightful throne. Earlier in his life, Horus was shown sitting in his mother's lap just as Jesus was shown sitting with Mary. Ancient Egypt casts a long shadow: the holy mother and child was an Egyptian invention.'

END OF CIVILISATION

We were back on the boat in time for cold towels and colder lemon juices. A gentle breeze struck up along the river, cooling us down and shattering the Nile's silvered calm. Palm trees bowed before it, donkeys brayed along the shore, while several large sailing boats laden with green bananas made good going of it. Engineering feats such as the Aswan Dam and the Cairo–Aswan road and railway have robbed the Nile of its age-old traffic (in ancient Egypt, as carvings at Edfu Temple had shown us, even the gods had travelled by boat), but there was still plenty of local traffic along that stretch of the river.

Plenty too to distract the eye while lunch was being prepared: a wagtail settling on deck, horses being brought down to the river for a wash, the extraordinary luminous quality of the light in the beautiful dom palms, which became more common as we went south and, wherever there was land to be worked, signs of farming, the basis for the success of Egypt's ancient empire and for its subsequent conquest by the Greeks and Romans.

As the afternoon went by, after we had passed Kôm Ombo's temple – its ruined hypostyle hall hanging dramatically on a rise over a bend in the river – a feeling of extraordinary serenity spread over the boat. Occasionally we were stirred from our reverie on deck or in cabin by the passing of another boat or by the reappearance of fishermen. But for most of the ride we were entirely captivated by the peaceful view and the happy prospect of reaching Aswan.

Late in the afternoon, with everyone on deck – united again – and even the Complainer lulled into a contented silence, we rounded a wide bend and Aswan came into view. To ancient Egyptians, Aswan was the end of the civilised world and the place still has some of the atmosphere of a frontier town about it. The light and landscape were different, the river was bent and twisted by granite banks, with pale yellow sand piled up behind them, a reminder of the desert which stretched from here the whole way across the continent to the Atlantic Ocean.

After we docked, I decided to do some shopping in the souk (market) and being unable to find a taxi that would take me there for a reasonable price, I accepted a ride on a donkey-cart. The sun was setting as we jogged along the corniche and people came out for a stroll along the river, a man knelt down amongst the scrubby bushes as the *muezzin* called the faithful to prayer, while the sails of feluccas fluttered in a breeze that carried with it a hint of the brown mud and rich scent of Africa. When the sky began to change colour, the river, chameleon-like, followed suit, turning from dark-metal to gold, then from blood-red to mercury and black, reminding me again that Egypt is still, as it always has been, the land of *al-Kam*, of alchemy.

OPPOSITE PAGE, *varied wares, not all Egyptian, are on display to lure tourists everywhere*

BELOW, *the observation deck on the commemorative tower on Aswan's High Dam is sometimes open*

PRACTICAL INFORMATION

■ Scores of boats make the journey between Luxor and Aswan and as standards of service, cleanliness, hygiene and comfort vary as much as their capacity (the number of beds varies from 40 to over 200), it is necessary to choose with care. As one would expect from the company that pioneered Nile cruising in the 1860s, Cook's boats are amongst the best. Usually only two nights are spent on the river between Luxor and Aswan, with a night or two on board at each end. Thomas Cook Holidays, Tel: 01733 332333, offer a range of tours from ten nights, including a stay in Cairo and a cruise between Luxor and Aswan on the *Nile Serenade* and its sister-ships, to a 14-day tour of Egypt.

■ Specialist operators include Soliman Travel, Tel: 0171 370 6446; Fax: 0171 835 1394 and Jasmin Tours, Tel: 01628 531121; Fax: 01628 529444. The Egyptian State Tourist Office, Egyptian House, 170 Piccadilly, London W1V 9DD, Tel: 0171 493 5282; Fax: 0171 408 0295, has details of other operators.

■ Islamic militants fired on boats between Cairo and Luxor several times between 1992 and 1995, but there have been no such incidents between Luxor and Aswan, where you are more likely to fall victim to the 'Pharaoh's curse', an unpleasant stomach upset.

■ An alternative to these floating hotels is to hire a felucca, an open sailing boat. The journey can take four or five days, sleeping out can be cold and not all *felucciyas* cook well, but it is a unique experience (information from the Luxor or Aswan tourist offices). Be sure to start in Aswan, so that if the wind fails the current will carry you down.

ABOVE, *floating in the air ... getting a bird's-eye view of African wildlife*

Wildlife Watch: from Kenya to the Seychelles

SHIRLEY LINDE

This exciting trip started with a three-day safari in the Masai Mara Game Reserve in Africa, then continued aboard the *Renaissance IV* for a 12-day cruise to Lamu, Zanzibar, Madagascar and the Seychelles Islands. In all the ports, the mix of various races and cultures – African, Arab, Indian, Chinese, Portuguese, Turkish, British, French – showed up in an exciting medley of architecture, music, dress and customs. To cap it all, we saw exotic flora and fauna that can be found nowhere else on earth.

It was 5am when we were awakened by a call at the tent door and before long found ourselves bouncing along a rutted dirt road in a Land Rover through the black night. Hyenas called out of the darkness. Scared by the headlights, a zebra bolted across the road, then a graceful gazelle bounded away on the left. Then an ugly warthog glowered at us as we rattled by. We were on our way to a hot-air balloon ride and arrived at our destination just before dawn.

As we climbed aboard and soared off, the balloon seemed huge and the heat and noise was more than we had expected. Gradually, however, white knuckles relaxed as we sailed over the grasslands and began spotting wildlife: a herd of elephants; a pride of lions; three ostriches running, ungainly, over the plain. Later, a rare black rhinoceros and her baby; then warthogs, hyenas, giraffes, zebra, wildebeests, gazelles. To crown it all, we enjoyed a champagne breakfast on landing.

BELOW, *zebras are one of the most commonly seen species on the Masai Mara Game Reserve*

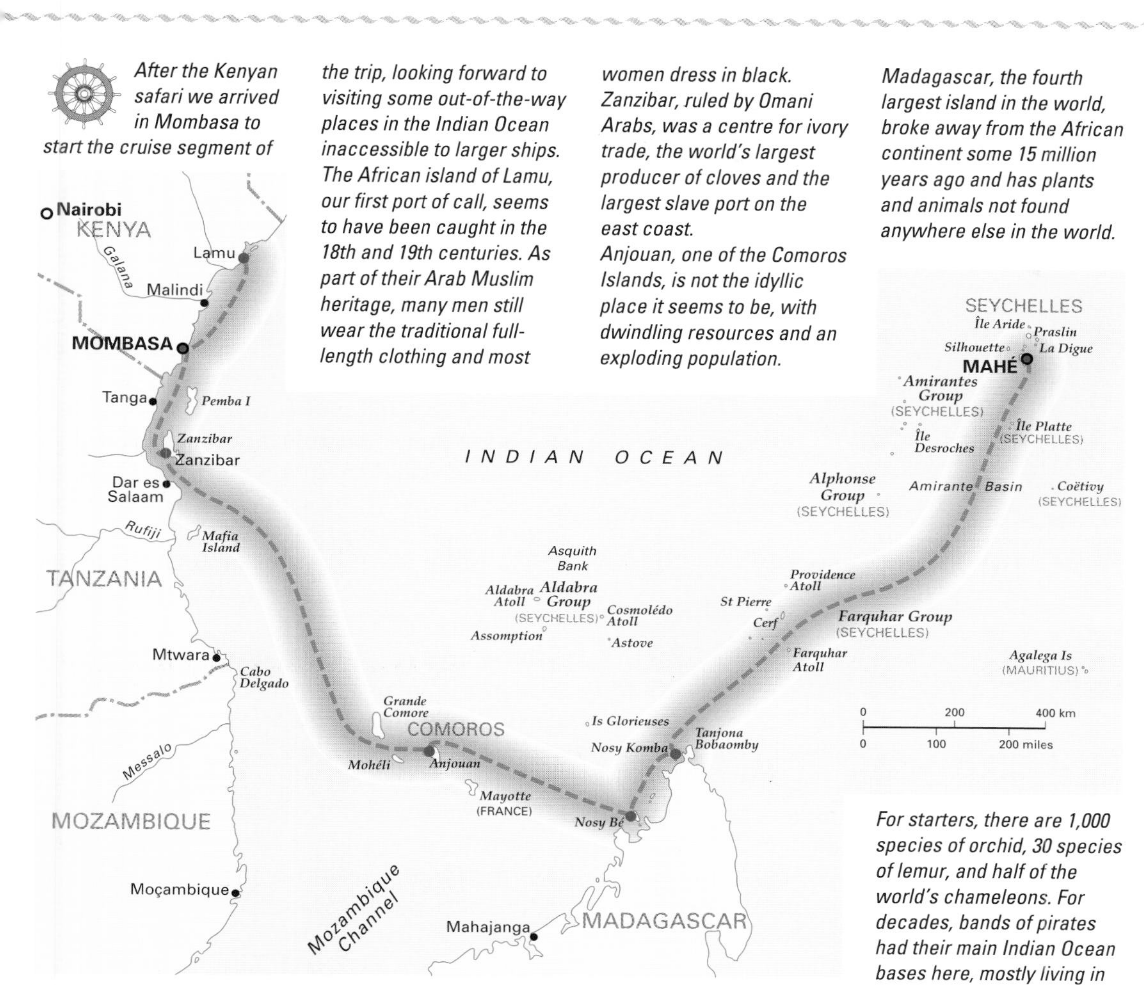

After the Kenyan safari we arrived in Mombasa to start the cruise segment of the trip, looking forward to visiting some out-of-the-way places in the Indian Ocean inaccessible to larger ships. The African island of Lamu, our first port of call, seems to have been caught in the 18th and 19th centuries. As part of their Arab Muslim heritage, many men still wear the traditional full-length clothing and most women dress in black. Zanzibar, ruled by Omani Arabs, was a centre for ivory trade, the world's largest producer of cloves and the largest slave port on the east coast.
Anjouan, one of the Comoros Islands, is not the idyllic place it seems to be, with dwindling resources and an exploding population.
Madagascar, the fourth largest island in the world, broke away from the African continent some 15 million years ago and has plants and animals not found anywhere else in the world. For starters, there are 1,000 species of orchid, 30 species of lemur, and half of the world's chameleons. For decades, bands of pirates had their main Indian Ocean bases here, mostly living in Libertalia. Pagan rites still take place, such as turning the dead and taking relatives from their graves during certain celebrations. As in Anjouan, the birthrate is exploding, deforestation is rampant, and much of the island's topsoil has washed away.
Finally we reached the magical Seychelles. These paradise islands are really the peaks of the submerged mountainous continent formerly joining Africa and India, now covered by water.

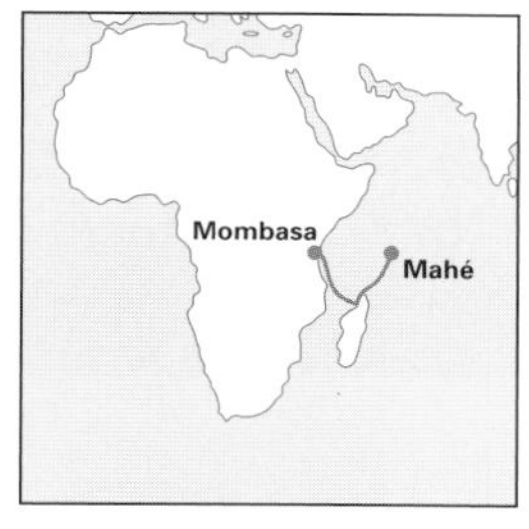

The balloon trip was one of the options offered by Micato Safari. We had set off after a night in Nairobi, flying in a 1942 DC3 over the Great Rift Valley to the 700-square mile (1,778sq km) Masai Mara Game Reserve, land of the Masai tribes just north of the Serengeti Plains and reputed to offer some of the finest game viewing in the world.

The plane landed on a grass airstrip in the middle of nowhere and we immediately embarked on our first game run, travelling in six-passenger vans with open roofs for photography. On our way to the camp we saw hundreds of zebra, giraffes munching from the treetops, cape buffalo and dozens of families of baboons. We were told the baboons frequently get tipsy on fermented fruits from the sausage tree, which is also used to make a local beer.

Masai tribesmen and women wearing traditional dress could be seen walking the roads or tending their herds. We learned our first Swahili word, *jambo* (meaning hello), from our driver and we heard it, given with smiles, wherever we went.

SEAWARD BOUND

After the safari we travelled via plane and bus to the seaport town of Mombasa in Kenya to meet the *Renaissance,* all looking forward to our exotic ports of call and the ease of shipboard life. The *Renaissance*, like her seven sisters on the line, is some 295ft (90m) long and 50ft (15m) wide with a cruising speed of about 15 knots. With only about 100 passengers on board, there is an inti-

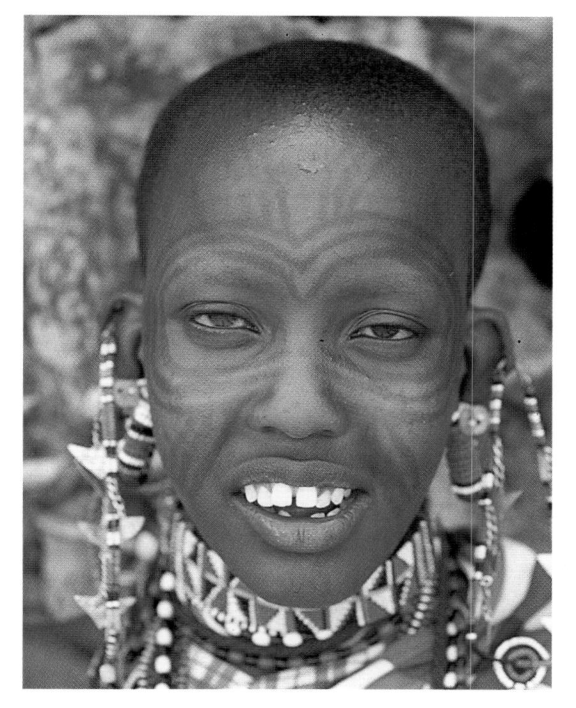

LEFT, *traditional Masai costume is worn daily*

ABOVE, *first stop Lamu; traditional dhows are still used for daily transport*

BELOW, *hustle and bustle in a Zanzibar market*

mate atmosphere and it doesn't take long for everyone to become acquainted. All the cabins face outside, have large windows and include a comfortable sitting area. Home comforts such as a telephone, television, refrigerated bar and bathroom with shower are standard, and passengers have a choice of queen or twin beds. Some of the staterooms have a private balcony. Nearly all the sailing is done at night which allows plenty of time for sightseeing during the day when the ship is usually docked.

Our first port of call was Lamu, an island just off the Kenyan coast. Here donkeys are the main mode of transportation and dhows (traditional Arabian boats with triangular-sail rigging) are used to get from island to island. The architecture in the narrow winding streets combines European, Arabian and Persian influences, with many buildings featuring heavy, ornately carved doors: they are also very practical – kitchens are upstairs because of heat and smoke, doors face only north or south as protection from the sun, there are large steps to keep children from going upstairs to adult bedrooms, and fish are kept in the bathing water to eat mosquito larvae.

The island has 31 Muslim mosques in all, the oldest of which was built in the late 1300s, and a fort, two centuries old.

I managed to get lost at one point, becoming separated from my group in the crowded narrow alleys filled with merchants, villagers, workmen and donkeys. But it was okay – I walked into a store to catch my breath and my bearings and found some wonderful, brightly coloured fabrics to bring home – red from Mangrove bark, black from ebony roots, white from shells.

Our next stop was Zanzibar, for centuries the commercial centre of east Africa and the most important town on the east African coast. A walking tour of the town revealed ornate old houses,

decorative wrought-iron work and carved teak doors among what are mostly neglected and rundown buildings. It was heart-wrenching to see the Sultan's Palace (it is rumoured that after the palace was built he killed the workmen or cut off their hands so that nothing like it could ever be built again), the former slave market with its whipping post, Livingstone House where the explorer Henry Livingstone lived before his last expedition, and the church with the crucifix made from the tree under which he died, in such a state of disrepair.

Out of Africa

Once back on board, and in between ports, there was always plenty to keep passengers entertained, from a piano bar and dancing in the main lounge to lectures on the history, architecture and culture of each place we visited. Other facilities included a library, a swimming pool, a casino, a gift shop, a hairdresser, a masseuse and a laundry. When at anchor, a water-sports platform was lowered from the stern, allowing access to the water for swimming, water-skiing or sailing. Snorkelling and scuba-diving were also there for the taking.

Anjouan is one of the Comoros Islands, and with its palm trees, beaches, rivers, a mile-high mountain, lakes and waterfalls it has all the appearance of a South Pacific gem; in reality, however, the island is set on a self-destruct course. Already overpopulated, the accelerating birthrate is expected to be double that of 1980 by the year 2000. Indiscriminate tree felling is causing soil erosion into the sea which in turn destroys the coral as it settles, resulting in fewer and fewer fish to catch and eat; schools and hospitals are on strike; and there is just one physician for every 14,000 people.

The best way to see the island is by the local wooden-bench minibus that careens around the hills and curves, while children line the streets, waving and shouting 'bonjour' as you whiz past.

Good buys include the local perfume, called *ylang-ylang*, and vanilla, the two major exports. In the town of Maatsamudu, the capital, you will see 17th-century houses, the market place where charcoal is sold and Muslim women with yellow mudpacks on their faces for the purpose of keeping the skin soft for their husbands.

Unique Island Wildlife

On reaching Madagascar our first stop was Nosy Bé. Again, we saw some wonderful old architecture but it was run-down and in disrepair. There were excellent carvings to buy, and hand-embroidered table cloths and shirts, plus a variety

ABOVE, *the harbour and mosque of Grand Comore, one of the Comoros Islands*

BELOW, *a female black lemur from the colony on Nosy Komba*

PRACTICAL INFORMATION

■ Cruise/safari trips are a relatively new combination holiday that allow you to sample several experiences as part of one trip – the photo safari in Africa, then on to some of the world's least visited and most exotic ports.

■ Pack a separate soft-sided bag just for the safari with casual wear only; your cruise suitcase will be stored in Nairobi for later pick-up.

■ Renaissance Cruises now schedule safari/Seychelles cruises in December, January and February, with Abercrombie and Kent Safaris. The trip does not now include Lamu, Zanzibar, Madagascar or the Comoros Islands; after the safari, passengers fly directly from Nairobi to the Seychelles Islands, boarding the ship at Mahé, the capital, and visiting the islands of Praslin, La Digue, Curieuse and Desroches.

■ SilverSea Cruises, the Cunard Line and Seabourn Cruise Line combine cruises with Micato Safaris. Of these, Cunard's *Sea Goddess I* and *Sea Goddess II* in November, December and January cruise from Mombasa or Mahé to various ports.

■ Silver Sea Cruises, with their 295-passenger sister ships *Silver Cloud* and *Silver Wind*, depart from Mahé and include Nosy Bé, Zanzibar and Mombasa.

■ In November, the 204-passenger *Seabourn Spirit* combines a safari with a cruise from Mombasa to various ports.

■ Renaissance Cruises, Tel: 800 525 5350 or 954 463 0982.
Cunard, Tel: 800 221 4770 or 212 880 7500.
Seabourn Cruise Line,

of stitched-and-cut-out linens unique to this area. Don't buy turquoise or ivory, however, as taking either out of the country is illegal.

That evening, the children of Nosy Bé came aboard in costume and presented us with a show of local folk dance and song. Delightful.

The second Madagascar stop was Nosy Komba, home to both a quiet, dignified village people with many crafts for sale, and a not-so-dignified colony of black lemurs, nevertheless protected by the villagers who hold them sacred. Lemurs eat early in the morning and sleep at midday so we were on the island by 8am, complete with bananas. 'Wear old clothes', our cruise director had said. It was good advice. The lemurs came hurtling out of the trees, thudding on to our backs and shoulders, sometimes on to our heads, jumping from person to person, from tree to ground to tree, grabbing and gorging on our bananas the while. The cruise director also explained about the short digestive system of the lemur. 'Don't walk under the trees after 20 or 30 minutes', he advised. He was right about that, too.

We walked through the village, admired and sometimes purchased the villagers' carvings and hand-worked linens, and walked along the beach. One woman was teaching some children a song in English. We all left feeling the world was indeed a wonderful place.

ENCHANTED ISLANDS

Epitomising the exotic plants and animals that characterised this trip were those found in the Vallée de Mai on Praslin (pronounced 'Praeleen'), our first Seychelles Island. Here the forest is

BELOW, *Victoria Harbour, Mahé*

hushed and majestic. Coco-de-mer palms, growing in only two or three known places in the world, soar 100ft (30m) above, their giant fronds making a patterned roof against the sky; some are 900 years old. They are quite sexually explicit in appearance, the dangling pistil of the male tree very evident and the heavy seeds of the female tree unquestionably a giant version of human female parts. The screech of a rare black parrot interrupted our reverie and we were told there are only 26 such birds left on earth. We left reluctantly, wishing we could transform some of the world back to its primeval forest state.

Could anything be better than the Vallée de Mai? Well, perhaps La Digue. I've seen many beaches on my journeys, but the beaches here are like no other. Huge granite boulders stand out against white sand beaches and palm trees growing

from crevices and cliffs overhang hidden coves. Transport is by bicycle or oxcart (the island has no taxis) and you can stop to see the giant tortoises, photograph the island houses with their pastel colours and ornate grillwork, visit the old cemetery, climb to the top of the island, or snorkel and swim at one of the secluded beaches. We did it all.

Aride, run by the Society for the Promotion of Nature Conservation, is home to thousands of rare seabirds – and seven human residents. Fortunately the sea was calm enough for us to go ashore by Zodiac and we saw fairy terns, frigates and noddys, plus a few roseate terns and, of course, the ubiquitous little iguana lizard.

Having travelled some 1,900 miles (3,040km) since leaving Mombasa we had reached Mahé, the end of our cruise and only four degrees from the Equator. Understandably, many people stay on for a day or two to explore the island's wonderful coastline, visit the tea, vanilla and cinnamon plantations or shop in the capital of Victoria, at the craft village where local craftsmen and women embroider, weave and construct model boats, or at the Seupot Co-operative where potters make original designs from local clay. A few of us chose to walk the Morne Blanc trail in the Morne Seychollois National Park, and were rewarded with a magnificent view over the island

Alternatively, why not take a half day, or even a week or two, to see the wonderful underwater world of the Seychelles. There are at least 200 species of fish and 150 species of coral around the islands and many boats and guides to help you find the best spots to see them.

And you might wish to eat of the breadfruit while you are there. If you do, it is said, you will be sure to return.

Tel: 800 929 9595 or 415 391 7444.
SilverSea Cruises, Tel: 800 722 6655 or 954 522 4477.
In Australia (all Sydney numbers): Cunard, Tel: 2 9956 7777; Wentworth Travel, Tel: 2 9362 4888; Creative Cruising, Tel: 2 9699 9199; Cruise World, Tel: 2 9966 1677; The Cruise Brokers, Tel: 2 9281 9543; Wiltrans, Tel: 2 9255 0899.

■ During the day temperatures are high, but take warm clothing too as air-conditioned ships are sometimes over-cooled in the evenings and safari country is cool in the mornings and evenings.

■ Visas are required for Kenya, Madagascar and Tanzania, plus some other ports of call.

■ Anti-malaria tablets should be taken and inoculations against yellow fever and cholera are required. Note that requirements sometimes change, so consult your local public health service for recommendations in good time before leaving.

BELOW, *wonderfully sculpted rocks on the island of La Digue*

Island Lifeline: the Royal Mail Ship to St Helena

ANGELA WIGGLESWORTH

ABOVE, *this passenger won't forget 'crossing the line' on the RMS* St Helena

BELOW, *Cape Town's waterfront, with Table Mountain as its unmistakable backdrop*

The 6,138-ton RMS *St Helena*, known affectionately as RMS (Royal Mail Ship), is the only scheduled long-distance mail service liner left in the world. She sails from Cardiff four times a year, and ten times from Cape Town, to carry passengers, cargo and mail to the island of St Helena, stopping at Tenerife and Ascension on the way. Once a year she delivers supplies to Tristan da Cunha, deep in the South Atlantic, the most isolated inhabited island in the world.

We drove through the early afternoon Cape Town traffic to the docks, and to 'J' berth. There she was, the RMS *St Helena*, with her bright yellow funnel and orange lifeboats. 'Welcome aboard', said Geoff Shallcross, the purser. 'Someone will show you where to go.' This turned out to be Pat, a young St Helenian, who took me to the comfortable cabin that was to be my home for the next 18 days.

At 5.24pm, to the sounds of *Hearts of Oak, Rule Britannia* and a bagpipe-playing passenger piping us off, the ship slowly moved away from her berth and we were off to face the long Cape rollers on our 1,557-nautical mile voyage south-west to Tristan da Cunha. It was to be five days before we saw land again. In the main lounge, Eddie was tinkling on his keyboard, groups were chatting and a very British tea was being served, with triangular sandwiches and a tin of biscuits to dip into.

THE ROYAL MAIL SHIP

The RMS is 345ft (105m) long and 62ft (19m) wide; she has twin engines, twin screws and stabilisers. Her service speed is 15 knots; she can take 128 passengers and 53 officers and crew, plus 2,000 tons of cargo. Everything on St Helena today (less than 17 years old) would have been brought by her or her predecessor: cranes, satellite dishes, horses. Once they had crocodiles on board.

The RMS, built in Aberdeen, was launched in 1989 by HRH Prince Andrew and made her maiden voyage a year later. She is managed by the Curnow Shipping Ltd in Cornwall on behalf of the St Helena Line. Her predecessor, the first RMS St Helena, *was requisitioned for the Falklands War; in 1986 it was decided to replace her with the present vessel.*
Tristan da Cunha, 1,500 miles (2,400km) west of Cape Town, is a roughly circular island 8 miles (13km) in diameter. A British Dependent Territory, with an administrator and elected council, the island can only be reached by sea and the RMS is one of only two regular cargo ships to dock here, though fishing vessels from Cape Town call six times a year.
Discovered in 1506 by Admiral Tristao da Cunha, the first settlers arrived here in 1816. It was in 1961 that the volcano erupted and islanders had to be evacuated to Britain for two years. Almost all chose to return. Today, there are 290 inhabitants, lobster fishing being their main source of revenue.
St Helena, 10.5 by 6 miles (17km by 10km), lies 1,694 miles (2,726km) north-west of Cape Town. It was discovered by the Portuguese in 1502 and became a British colony in 1659. Napoleon was exiled here after the Battle of Waterloo until his death in 1821; during the Boer War, 6,000 prisoners were brought here.
Before the opening of the Suez Canal in 1869, over 100 ships a year visited St Helena en route for India; today the island can still only be reached by boat. It is a British Dependent Territory with two dependencies of its own, Ascension and Tristan da Cunha, a Governor appointed from London, and an elected council. The population is 5,500 and their ancestors, all brought by the British, have come from Africa, Europe, China and Asia.

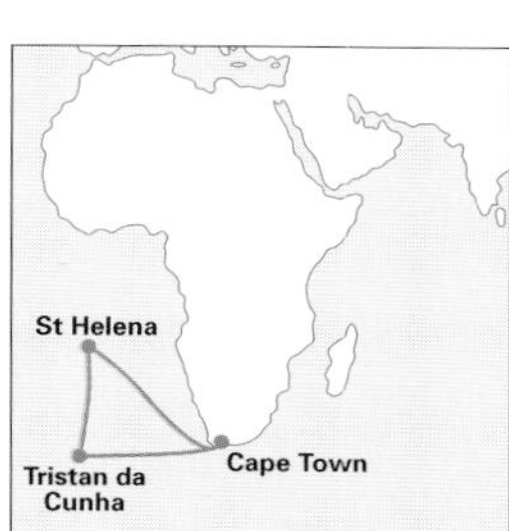

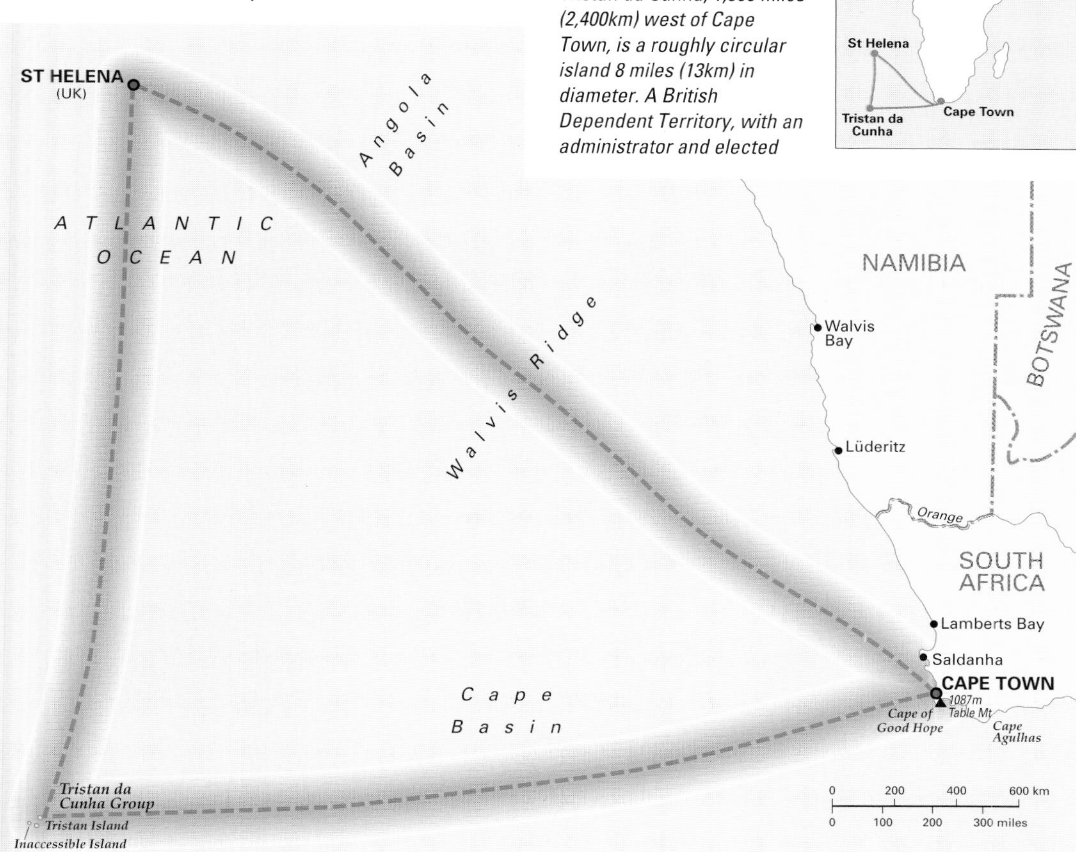

ABOVE, smart and workmanlike, RMS St Helena *on the Waters of Ascension*

ABOVE, *the snow-capped island of Tristan da Cunha disappearing below the horizon as we journeyed on*

The ship has two lounges, two bars, a dining room, a sun deck with swimming pool, a shop, a book and video library, a small gym, a hospital, a do-it-yourself laundry and a playroom. The *Ocean Mail*, slipped under the cabin door each morning with an early cup of tea, informed us whether dress for dinner that night would be formal or informal and listed the day's events. These included lectures, films, taped concerts, keep-fit sessions and deck tournaments, plus dancing, racing and quiz nights and a Welcome Aboard party. You could also visit the engine room, the galley and go to the bridge whenever you wanted.

One breathtakingly still night, when the sea seemed quite motionless, there was an excellent barbecue on deck followed by dancing under the stars and a moon that rose like a huge orange ball from the horizon.

With a friendly and helpful crew, and a small number of passengers made up of returning Saints (as the St Helenians are called), Germans, French, Americans, British and South Africans, one made easy friendships. We learnt much from the Saints about the complex problems facing their island and the unjust treatment (loss of full British citizenship and the right to work in the UK) that they were receiving from the British Government.

The days somehow blurred into each other as we sunbathed on deck, ate, read, swam in the pool and joined in the ship's entertainment. On the second day we were lucky enough to see the wondrous wandering albatross with its 13-ft (4m) wingspan, wheeling overhead, and the glistening dark-grey backs of right whales swimming near by.

On the third day, a 40,000-ton Japanese woodchip carrier appeared on the horizon. The weather on Tristan da Cunha was good, they reported over the radio. The omens seemed favourable for landing, though this was always difficult, as the captain warned in a pre-arrival briefing.

HOISTED ASHORE

The RMS can't go into the harbour, he told us; we anchor outside and passengers are taken ashore in small boats. If the sea is calm, the gangway can be used; if rough, it's the pilot rope ladder down the side of the ship, and a 4-ft (1m) jump into the boat below. As his audience went suddenly quiet, he said: 'There is no such thing as calm water in Tristan. You need to be fleet of foot and conditions can deteriorate quickly. If an officer says you don't go, you don't go.' We were to reach the island early the following morning.

At 4.30am I was on deck. Was that a cloud on the horizon or Tristan da Cunha? It was Tristan –

BELOW, *longboats on Tristan – a world apart*

a solitary cone-topped rock in an endless sea. Giant southern petrels and sooty albatrosses skimmed the water around us as we got near enough to identify some of the island's intriguingly named places, such as 'the Hut Where the Goat Jumped Off', 'Where the Minister Left his Clothes', and so on.

Our anchor grated down, but on the bridge the captain, in radio contact with Tristan's administrator, Brendan Dalley, said conditions on the radar screen looked bad. There was a force 9 wind and he didn't think we'd be able to land that day. Dalley agreed. It was decided we should shelter at nearby Inaccessible Island until the weather improved. Waves were spraying high against the ship's windows as we breakfasted on eggs and bacon and hot coffee, cocooned in a tiny, warm island of our own.

Seven waterfalls, cream-coloured from guano, crashed spectacularly down 2,631-ft (800m) high cliffs that gave us shelter for the day. It rained all morning and the fishermen among us cast their lines into water thick with brown kelp; by evening, though, the sun shone with a white light over the dark cliffs.

A tiny motorboat appeared, bobbing over the sea towards us. It was from Tristan's off-shore fishing vessel, coming to collect some parcels. 'How long have you been at sea,' one of the men shouted up as he drew alongside. 'Five days', said the captain. 'What about you?' 'Two-and-a-half months', he replied, catching the packets thrown down to him, plus some cans of beer.

Shearwaters glittered like stars in the sunlight, great southern petrels floated on the water waiting for fish scraps from the crew's barbecue, and huge tuna fish porpoised from the water.

Next morning hopes of landing on Tristan again faded. The weather was poor, with a north-northwesterly wind force 7. Then chance took a hand. The British Antarctic Survey ship, HMS *Endurance,* was in the area and had offered to fly essential personnel and supplies to the island in its helicopter. We crowded the upper deck to watch the operation as the helicopter hovered overhead and cargo was winched up. Then a voice came over the loudspeaker: 'Would Angela Wigglesworth go to the purser's office.' 'You can go ashore,' Geoff told me in a 'you can go to the ball' kind of voice, but as I've always suffered from vertigo I didn't greet the idea of being winched up 60ft (18m) in a wind force 7 with Cinderella-like enthusiasm. However, I'd been planning this trip for a year and the island's administrator had arranged for Jimmy Glass, the chief islander, a descendant of one of the first British settlers, to show me around. How could I then refuse the invitation? Up I went in what looked like an orange space suit, quaking with fear, eyes tight shut, until the wonderful solidness of the helicopter floor was firmly beneath me.

ABOVE, *three of the seven waterfalls of Inaccessible Island crash their way down the lush green cliffs*

Tristan is a tiny community, 290 strong. The cottages are whitewashed with asbestos roofs, and black lava walls edge the neat gardens, while the dark mountain towers above them. There is a well-equipped hospital, two churches, a school with 41 children, a supermarket, a swimming pool, and a field where football-rounders is played – a game that can include as many people as want to play divided into two teams. There is also a post office, a council chamber, a café, a museum and a fish factory. The potato, a vital vegetable on the island, is grown in patches 2 miles (3km) away. We learnt that the Dong (a disused oxygen cylinder) was hit with a hammer to tell everyone it was a good day for fishing; how each married couple was allowed to graze seven sheep and two cows on settlement land, but as many as they like on the plateau above; and how the island managed its crayfish industry to prevent over-fishing. We heard how young islanders longed to have their right to work in the UK restored along with full British citizenship, and we felt angry and sad that these people, so loyally British, were denied these rights by the British Government. I could only glimpse a little of this friendly, caring community, but I felt privileged to have been there at all.

SAINTS RETURN

As we sailed north to St Helena, we were, the captain told me, at least 400 miles (640km) away from any registered shipping routes. There was no room, therefore, for error. He was lucky in his regular crew, he said, for they were all highly trained St Helenians whom he had known for many years.

Four days and a 1,315-mile (2,104km) journey later, we reached St Helena and anchored outside the harbour. Immigration officials came on board, and after we had paid £10 for a certificate of entry we boarded the small boats that had come to ferry us across to Jamestown, lying snugly in a steep-sided narrow valley. 'I can't believe this is really happening,' said one St Helenian, back for the first time in 46 years, and straining to see once-familiar places. 'It's a dream come true.'

Colin Corker was waiting on the quayside with his 1929 open-topped charabanc and 14 of us piled in for a sightseeing tour of the island. We drove through the castle arch and up the main street with its Georgian houses and little shops, past the market building and a few pubs, and out into a countryside of dramatic contrasts: volcanic ridges, deep, green valleys, and flax-covered mountains slopes. Until 1965, when the British Post Office substituted St Helena flax with a synthetic fibre for their string, flax had been the major industry here; now there was hope that coffee growing would take off.

We visited The Briars, where Napoleon spent the first few and seemingly only happy months of his stay on the island; Longwood House, where he lived until he died; and his black-railed, now empty tomb – all owned by France. White arum lilies and agapanthus grew in the fields, pink hibiscus and pale blue plumbago in the hedgerows, and jacaranda and thorn trees were in full flower.

We stopped at Plantation House, home to the island's governor along with the famous tortoise, Jonathan, thought to be over 150 years old. Next we peered down Ladder Hill where, 600ft (183m) below at the bottom of 699 steps, Jamestown looked like a toy town.

Some passengers slept on the ship for our two-day visit to the island, coming ashore each day by boat. I booked in at Wellington House, a delightful, quiet hotel in the main street, with good food and spacious bedrooms. You can't walk anywhere in St Helena without being greeted with a friendly 'hello', and I had the chance to meet several local people: there was George Benjamin, a modest and knowledgeable man, famous for rediscovering the endemic ebony flower thought to be extinct; Jessica March, who makes lace good enough to give to the Queen (she did), and Greta Musk, a magistrate and first woman sheriff.

Then it was back to the ship. The cargo had been delivered, the hatches closed, next stop Cape Town, 1,813 miles (2,901km) away. There was a faint sense of anti-climax – St Helenian friends we'd made had left the ship, and we'd seen the two islands we'd come to see. But there was still the South Atlantic Ashes cricket match to be played on deck; passengers beat officers – a sad day, said the captain. Near the town of Lüderitz, on Namibia's desert coast, we saw two diamond-mining vessels and we boomed a salute as we passed the lighthouse at Dias Point where the keeper, a Mrs Clay, was once an RMS passenger.

It was dark when we approached Cape Town and a necklace of coloured lights sparkled round the base of the floodlit Table Mountain. It was so beautiful we found ourselves speaking in whispers. The pilot boat come speeding out towards us and, in Errol Flynn style, the pilot jumped aboard. Two tugs escorted us into port and nudged us, imperceptibly, right next to the quayside. 'Stop the engines', the captain said. The engines stopped. The voyage to these isolated, friendly islands with their unique communities was over.

PRACTICAL INFORMATION

■ Tour operators: Curnow Shipping Ltd, The Shipyard, Porthleven, Helston, Cornwall, UK. Tel: 01326 563434; Fax: 01326 564347. Southern Africa Travel, I Pioneer Business Park, Amy Johnson Way, York Y03 4TN. Tel: 01904 692469. Strand Voyages, Charing Cross Shopping Concourse, Strand, London WC2N 4HZ. Tel: 0171 836 6363; Fax: 0171 497 0078.

■ The RMS sails to St Helena four times a year from Cardiff, and 10 times from Cape Town; once a year she goes to Tristan da Cunha. There's a variety of fly-cruise package tours (lasting on average three-and-a-half weeks) with a choice of sailing or flying between Cardiff and Cape Town.

■ The best time to go is November to February, the southern summer.

■ The ship has two-, three- and four-berth cabins with shower, wash basin and toilet, 28 budget berths and one cabin designed for wheelchair passengers.

■ On St Helena, there are two hotels – the Consulate and the Wellington – and self-catering accommodation is available.

■ British Sterling and St Helena currency (with the same value as Sterling) are used on board as well as travellers' cheques and credit cards. On St Helena, travellers' cheques can be changed, but credit cards are not accepted.

OPPOSITE PAGE, *Jamestown and the RMS*

LEFT, *Longwood House, Napoleon's final home on St Helena*

ABOVE, *the impressive Shwedagon Pagoda is more than 300ft (90m) high and is completely covered in gold*

The Road to Mandalay

MARY TISDALL

Throughout the spring and summer the luxury Orient Express ship *Road to Mandalay* cruises the calm waters of the River Ayeyarwady between Bagan and Mandalay. This sleek white vessel glides slowly past a dramatic eastern landscape of snowy white temples, golden-tipped pagodas and green paddy fields still ploughed by oxen. This is a voyage that drifts back to a time long before Rudyard Kipling penned his immortal words 'On the road to Mandalay, Where the flyin'-fishes play'. Yes, this river was his road, which flows past the dreamy Asian remains of a 2,500-year-old civilisation and the fascinating wonders of ancient Myanmar (Burma).

BELOW, *the gleaming white cruise ship* Road to Mandalay *anchored at the riverside*

Passengers on the cruise ship the *Road to Mandalay* usually start their holiday by staying for one or two nights at the Inya Lake Hotel in Yangon (formerly known as Rangoon), 4 miles (6km) north of the city centre. Set by the lake from which it takes its name, the hotel has extensive grounds with a large swimming pool surrounded by tropical trees and shrubs. At sun-

The Road to Mandalay *was built in Germany in 1964 as a Rhine cruise ship. It entered service in Myanmar (Burma) in December 1995 after a $6 million refurbishment.*
Myanmar lies between Thailand and Laos to the east and Bangladesh to the west, with India and China on the northern border. Its greatest length north to south is about 1,300 miles (2,090km), east to west around 574 miles (918km). Along the coastline are thousands of small islands, mostly uninhabited.
The River Ayeyarwady (Irrawaddy) flows about 1,350 miles (2,160km) from the Katchin Hills in the north to the Andaman Sea and the Bay of Bengal in the south. Although the river divides Myanmar in two, the Ava Bridge near Mandalay is the only bridge over it – evidence of the lack of development in the country. Bagan, in the middle of the country, is an arid zone with savannah-like growth, thorn scrub, cacti, rice and betal palms.
Nearer Mandalay, the river banks are more fertile and crops of onions, garlic, bean and maize are cultivated. In the distance the Shan Hills can be seen, and further east there are dense forests of hardwoods and wild animals. The flat plains seen so well from the ship are dotted with the pinnacles of pagodas and temples and monasteries, while small village settlements cling to the river banks. In the countryside, oxen, bullocks and pony carts are the main means of transport.
Myanmar's history is a story of strife and hardship, and this holds true up to the present day. Recently the goverment decided to attract foreign investment and gradually the door is being opened to overseas tourists.

ABOVE, *tranquil Lake Inya in Yangon is surrounded by a richly wooded park*

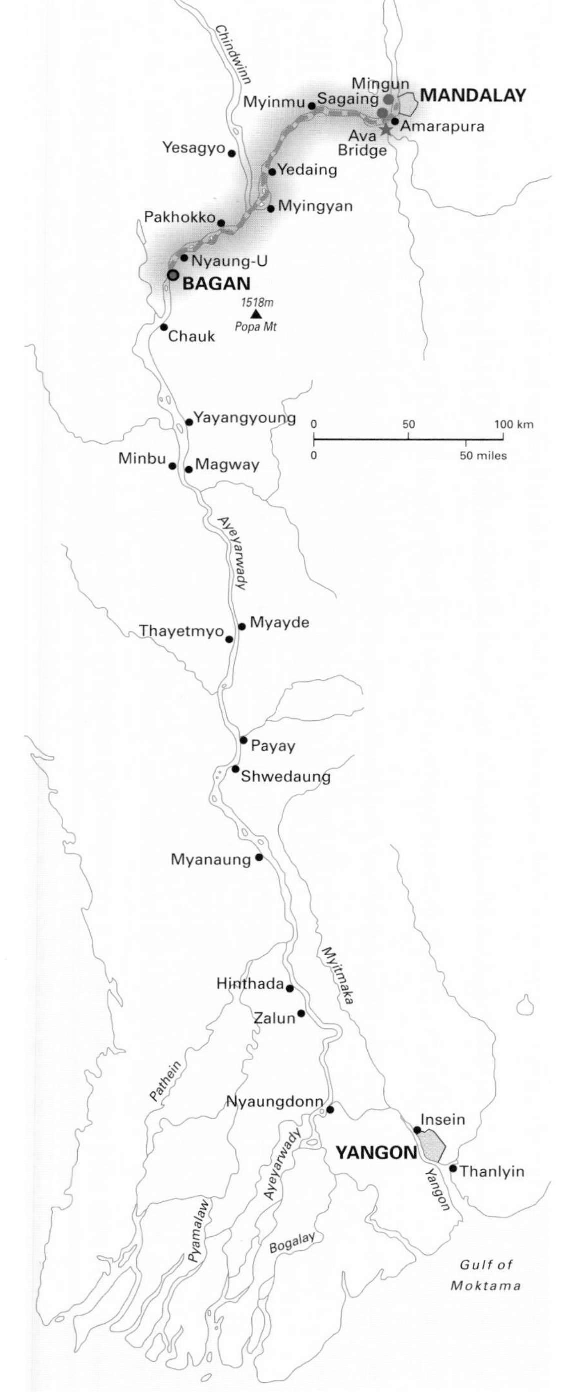

set, the peace is disturbed by the vibrant chatter of hundreds of birds coming to roost for the night.

I was introduced to some of my fellow cruise passengers, most of them wealthy Americans, by a charming dark-haired receptionist named Gi-Gi, whose family lived in the north of Myanmar. Together we decided to take the afternoon Yangon City Orientation tour, one of the optional coach excursions available prior to joining the ship.

Shore Excursions

A lively yet laid-back place, Yangon retains an air of the leisurely past. Many of the city streets are wide and lined with pink oleander bushes, tall palms, delicate blue jacaranda and red flame trees. The main roads are busy with ancient Japanese cars and overcrowded buses; bicycles and horse-drawn carts weave in and out to the constant honking of taxi horns; side roads are often dusty sand tracks. Downtown, by the busy Yangon River, fine old colonial buildings are now used as museums and foreign embassies.

The next morning I again joined a coach tour, this time to visit the Shwedagon Pagoda in Yangon, which, for all Myanmar's Buddhists, is the most sacred of sites. The main *stupa* (shrine), supposed to be 2,500 years old, has a brilliant 321-ft (96m) high golden orb encrusted with more than 4,000 diamonds, making it visible from almost everywhere in Yangon. Around the main shrine are 68 smaller pagodas, hundreds of pavilions and ornate temples, all adorned with flowers, statues and sacred relics. What impressed me most was the religious tranquillity that pervades this huge pagoda where vermillion-clad monks and cheerful families make various offerings and say prayers.

The following day we took a 40-minute flight with Air Mandalay north-east to Bagan, and after a short coach drive arrived on the muddy banks of the Ayeyarwady River. This was my first thrilling sight of the *Road to Mandalay* cruise ship, trim and white at anchor in the middle of the wide

ABOVE, *teak is one of Myanmar's main exports, and naturally travels by river*

river. After taking a photograph of local women doing their washing in the river, I scrambled down a sandy slope to board a small craft that ferried us to the ship.

On board we were greeted by Carl Henderson, the hotel manager, and a smiling multi-lingual crew from Myanmar, Thailand and Europe. On being shown to my cabin the first thing I noticed was the extra-large window between the twin bunk beds, great for viewing the river and beyond, and flowers were a welcoming touch.

Soon I was exploring the four main decks, mostly furnished with teak or rattan furniture made by local craftsmen. From the observation deck, with its pool and bar, the banks of the Ayeyarwady present an exotic picture of flat green rice fields, and in the distance the thatched roofs of village houses snuggle by tall trees. Throughout this voyage the pinnacles of pagodas glinted in the sunlight and tiny boats paddled past, crammed with people and supplies, often loaded with heavy teak logs.

BELOW, *typical local transport in Bagan – pony and trap*

The warm air sent me to the bar for a cool iced drink before descending to the restaurant to enjoy some tempting Burmese curry (spicy but not too hot) and a cold buffet luncheon. All the restaurant food, co-ordinated by the ship's Welsh chef, was consistantly good and attractively laid out.

After lunch it was time for a coach tour around the Bagan region. It is said that once there were over 4,000 religious sites in this area covering just 16 square miles (41sq km) and, despite wars, earthquakes, Kublai Khan and his Mongol force and constant pillaging, over 2,000 of these beautiful Buddhist buildings have survived.

Our tour began at the early 12th-century Ananda Pagoda where we gasped in wonder at the delicate murals and four colossal 31-ft (9.5m) high gilded Buddha images. Guides point out that if you stand by the donation box in front of the southern statue his face looks sad, while from a distance he looks happy. It is here, during the Festival of Pyatho, held in December and January, that up to 1,000 monks chant day and night.

Bagan town is the main centre of the region's lacquerware industry and we visited one of the factories. Young girls and boys sat on the dirt floor creating and applying the black lacquer to paper, cane or wooden household articles. It is a long and labourious task, taking up to to six months to complete a piece of furniture. Most designs are inspired by religous scripture and are carefully drawn freehand.

We also went inside a small house on stilts, which turned out to be a silverware factory. Here, again sitting in rows on the ground, young teenagers hammered intricate patterns on silver objects. Silverwork is widely available at quite reasonable prices but it is inadvisable to buy gold jewellery and precious stones from street vendors as they may be fakes; buy only from government-registered traders, found in some markets and most hotels. Myanmar's pigeon-blood-red rubies are said to be the finest in the world.

The cruise programme includes optional early morning excursions to see the sights, or an alternative is to hire a pony and trap to explore alone, at a later time. I chose to walk around the meat market, which requires a strong stomach, given the carcasses of fly-ridden freshly slaughtered animals, live chickens in baskets and trays of fried locusts and cockroachs for sale. However, the huge mounds of fruit and vegetables, mangoes and papaya, plus spice stalls, provided a colourful picture. Well-made cane baskets, lacquer bowls and trays and oriental puppets – often with as many as 60 strings and hand-embroidered clothes – all make worthwhile souvenirs. The local Mandalay rum is quite palatable, too!

ABOVE, *lacquerware is painstakingly handcrafted, often in cramped conditions*

RIVER LIFE

It was mid-morning when our voyage finally began; how good it was to be moving at last on the wide, fast-flowing river. Captain Ba Nyan, a Thai, explained to me that the river presents a continual navigational challenge as the flow of water differs dramatically from day to day, with sand barriers shifting within hours. Where the river narrows, the current accelerates and whirlpools sometimes appear. Fortunately the ship is fitted with an electronic depth sounder, but there are still drifting teak logs to contend with and the volume of river craft requires constant vigilance.

Along the river sandbars form little islands where itinerant fishermen make temporary bamboo homes and ferry their catch to market in frail little canoes, often using a single oar against the strong currents. Once we saw a boat being pulled with a long rope held by two small boys walking along the bank of the river while the mother shouted instructions from aboard! Sometimes these small boats are filled with huge mounds of fine-quality sand, sold up and down the river for building work.

Rural life in the quiet villages is simple, having changed little over the centuries, and the folk are practically self-sufficient. They grow vegetables, keep poultry and eat rice and fish, which is often cooked on a wood fire; washing and bathing is still done in the muddy river.

Life on board ship is friendly and easy. On the first evening of our voyage

LEFT, *Captain Ba Nyan of the* Road to Mandalay

BELOW, *ferry boats are just one of the many types of craft plying the Ayeyarwady*

RIGHT, *the white* stupas *of Sagaing are a glorious sight on the approach to Mandalay*

the ship moored on the river and a cocktail party was held on the observation deck at which officers and members of the 80-odd crew from Myanmar, Thailand and Europe met the guests. Later, the merriment continued up on deck under the stars.

REACHING MANDALAY

The next day we were on the move again. Just after lunch, near Mandalay, we sailed under the central arch of the graceful Ava Bridge. Built by the British in 1934 and rebuilt after the war, it is the only bridge to span the mighty Ayeyarwady River. The *Road to Mandalay* has a private mooring just below Sagaing Hill, and as the ship approaches the jetty there is a fine view of a line of white *stupas* (shrines) dotting the hillside.

Shortly after the ship berths a coach leaves for a tour of Mandalay. The city, the last Burmese capital before the British took possession in 1886, was severely damaged at the end of World War II while being liberated from the Japanese. Despite this, however, there is much to see, especially the great walls that once surrounded King Mindon's Palace and a beautiful lake where delicate, pale pink lotus flowers float on the shimmering water.

Above the city Mandalay Hill rises to 774ft (235m). Should you think of walking to the top, bear in mind that there are 1,729 steep steps involved. We reached the top of the twisting, narrow road by a hair-raising coach ride – the driver's assistant stood on the step by the open door with a large wooden block to use as a wheel chock in the event of the bus having to make an emergency stop! The views from the summit are breathtakingly spectacular, particularly as the sun

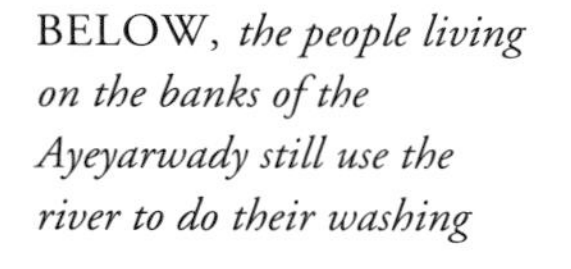

BELOW, *the people living on the banks of the Ayeyarwady still use the river to do their washing*

disappears over the horizon. Below is the sprawling city and beyond, towards Sagaing, the white *stupas* and golden-tipped pagodas stand tall and proud, guardians of the country's long and fascinating history.

Back on board after a dinner served by smiling Burmese waiters and pretty *longyi*-dressed waitresses, it was time for everyone to go to the observation lounge to watch a presentation of tribal costume, during which heavily embroidered, brightly coloured clothes from various regions of Burma were displayed by shy young models. Other evening entertainment aboard included folklore music and an animated Bagan puppet show.

On our last full day of the cruise an excursion to Mingun provided an opportunity to see river and rural life at close quarters. A local ferry took us upriver to the village, on the northern bank of the Ayeyarwady, 7.5 miles (12km) north of Mandalay. Mingun has two notable sights, the Pahtodawgyi Pagoda and the Mingun Bell, the world's heaviest uncracked bell, weighing around 92 tonnes. Visitors are invited to strike it for good luck.

At the end of the cruise many of us agreed that our brightest memory was watching the Offering of Alms to a procession of monks, held adjacent to the *Road to Mandalay* mooring on our last day. A table of food was taken from the ship and put on the quayside; a few moments later a solemn line of about 20 vermillion-robed young novices and monks, whose ages ranged from seven to 90 years, walked in slow, silent procession carrying their lacquer begging bowls (they have to live by alms alone). With much reverence, members of the ship's crew served the monks with food then bowed low in respect. This was done with great dignity and with no heed to the clicking cameras.

Myanmar is a remarkable land, locked in a time warp, and travelling along the Ayeyarwady in style is a remarkable voyage – and as I did not see those flyin'-fishes, I have an excellent reason to return.

LEFT, *the Mingun Bell is one of the largest ringing bells in the world*

PRACTICAL INFORMATION

- All cabins have en suite shower and WC facilities, personal safe, telephone, satellite television, towels and toiletries. All meals are included.
- The climate is monsoonal, with the dry season being November to April.
- *Road to Mandalay* river cruises (3, 4 or 7 nights inclusive packages) can include hotels, flights and the Orient Express train journey from Singapore to Bangkok .
- Head Office: Orient Express Trains and Cruise, Sea Containers House, 20 Upper Ground, London SE1 9PF. Tel: 0171 805 5100; Fax: 0171 805 590.
- International Reservations:
Australia Tel: 2 232 7499.
Belgium Tel: 2 223 2423.
France Tel: 1 45 62 00 69.
Germany/Austria Tel: 211 162106/7.
Italy Tel: 2 5518 0003.
Japan Tel: 3 3265 1200.
Korea Tel: 2 755 9696.
Netherlands Tel: 35 695 5111.
New Zealand Tel: 9 379 3708.
Singapore Tel: 323 4390.
South Africa Tel: 11 886 7270.
Spain Tel: 3 215 2494.
Switzerland Tel: 22 366 42 22.
Taiwan Tel: 2 567 3081.
USA Tel: 800 524 2420.
- Royal Brunei Airlines, 48 Cromwell Road, London SW7 2ED. Tel: 0171 584 6660; Fax: 0171 581 9279.
- Return flights London to Yangon direct – twice weekly

A Slow Boat along the Yangtze River

CHRISTOPHER KNOWLES

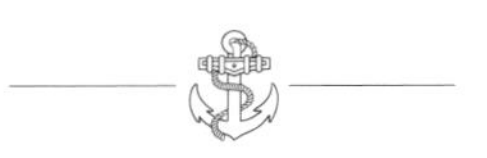

The Yangtze cruise provides a wonderful way to discover one of the world's great rivers and at the same time obtain an insight into a changing China. Now it is possible to make the journey in considerable comfort, but it is also possible to travel more cheaply, and in the way that most Chinese travel – by using one of the standard ferry boats which ply the river between Chongqing and Shanghai. In the following account the journey takes place in the first years after China began to welcome foreigners, but whilst startling changes have taken place in China as a whole, the essentials of the voyage remain unaltered.

To get to Chongqing, where the journey in this account begins, would involve a flight from within China. The city itself, which according to some reckonings is the largest in China in terms of population, is in the province of Sichuan, bordering Tibet in the far west of China. It is built on hilly land at the confluence of the Jialing and the Yangtze, a bustling, crowded port in a province that is among the most prosperous in the new China. The casual visitor to the city would perhaps enjoy a cable-car ride across the river, or a visit to the enchanting Buddhist grottoes at Dazu, which is a half-day drive from the city. Otherwise, just eat the delicious spicy Sichuanese food which is a highlight of any visit to the region.
The Yangtze – or Chang Jiang as it is more correctly known – is 3,938 miles (6,300km) long, the third longest river in the world. It rises on the Tibetan Plateau and disgorges into the Yellow Sea. Joined by some 700 tributaries, it may be divided into three sections: the Upper Reaches from its source to Yichang – this is the most spectacular part which includes the Three Gorges; the sluggish Middle Reaches between Yichang and Hukou; and the Lower Reaches from Hukou to the mouth, an area known as the Land of Fish and Rice.
Before reaching the gorges, the Yangtze passes a number of places of interest – the red pagoda at Shibaozhai; small riverine towns such as Fuling and Wanxian with their markets; rock formations and the legends attached to them. The gorges themselves are associated with a plethora of legends, many based on natural rock sculpture. Nowadays the majority of boats stop after the first gorge so that an excursion can be made to the Little Three Gorges along the Daning River.
After the gorges, the boats pass through the Gezhouba Dam and then travel on to Yichang, where the White Horse Cave is well known for its stalactites and stalagmites. Beyond here the river is slow and wide. It passes the old foreign concession city of Wuhan, an important industrial centre with an excellent museum, a carpet factory and an ancient temple.
The full route then crosses the province of Jiangsu, with its canals and paddy fields, to the walled city of Nanjing, and finally to Shanghai and the Yellow Sea.

In China there is one singular paradox which is a foil to all who try to make sense of her. It is that in China nothing changes and that simultaneously nothing stands still. Perhaps, as China 'reconstructs', the paradox will resolve itself, and that long legacy of timeless images will be consigned to coffee-table photography books; but that is still some way in the future.

In the meantime the daily grind for the greater part of the population goes on much as it has done for hundreds, if not thousands, of years, albeit without the long shadow of the Cultural Revolution hanging over it. In the great cities there are

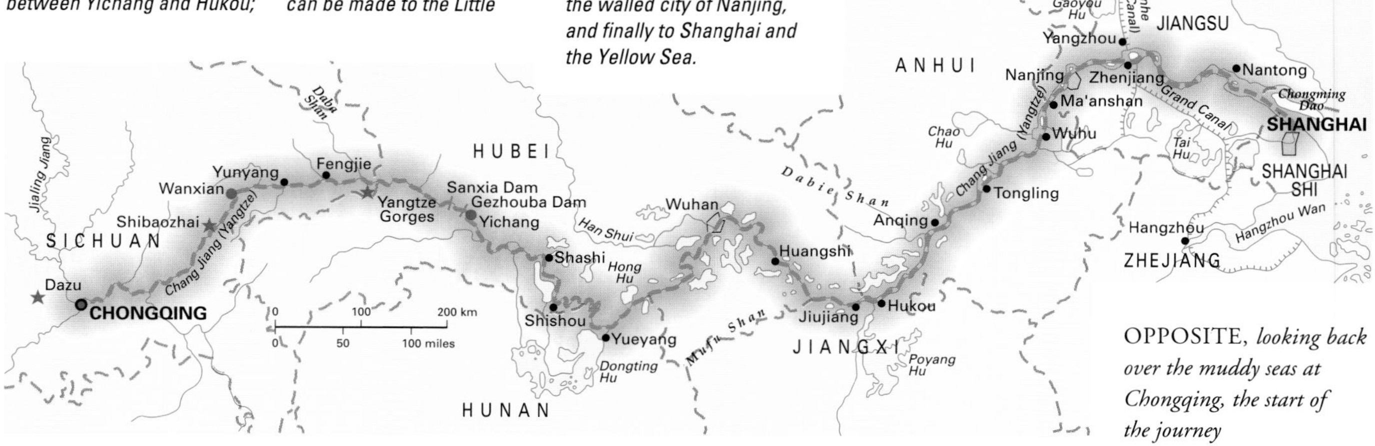

OPPOSITE, *looking back over the muddy seas at Chongqing, the start of the journey*

ABOVE, *a cable-car ride above the rooftops of Chongqing: the city is built on steep hills*

BELOW, *passenger cruisers on the muddy Yangtze*

shops overflowing with imported goods and the spirit of free enterprise has caught on with sometimes alarming ferocity. In the enormous spaces between these centres, along the banks of the Yangtze, for example, things are a little slower to fall into line.

It is true that now anyone with enough money can cruise along the Yangtze in considerable comfort, in cabins with air-conditioning and showers, and in the heat of a Chinese summer such luxury is not to be lightly dismissed. But anyone interested in getting under the skin of the country probably needs, within reason, to do what the average Chinese does. For a start few, if any, Chinese would be on the river for pure pleasure – they would be moving from place to place for some practical purpose connected with work or with family. They would, furthermore, be berthed in one of the many classes of cabin in one of the dozens of ferry boats that routinely ply the river between Chongqing in the west and Shanghai in the east, stopping at all ports between.

HEAT AND CHAOS

Chongqing, situated at the hilly confluence of the Jialing and the Yangtze, famous because it is about the only city in China without hordes of bicycles, is by no means beautiful. Known as one of the Three Furnaces of China because of the fierce heat that builds up in the summer months, it is every inch a port and therein lies its main appeal.

The ferry boats line up against the quay which is reached down a long flight of steps up and down which pad an endless procession of porters and passengers. The Chinese seem to be incapable of travelling light. Everyone struggles down with an almighty load, black plastic suitcases threatening to split, red, white and blue synthetic hemp bags stuffed with food, and bundles of fruit in string bags or tied together with bits of raffia. Of course many of the local people are immensely poor so

they bring all the food that they are likely to need for the journey.

The river is brown, dense with mud, glinting brassily in the early morning sun which is already uncomfortably hot. The ferry boats, painted sea green and off-white, all bear the same name, *East is Red* (after the communist anthem favoured by Chairman Mao), which is inscribed in large black Chinese characters. From the exterior these river-craft look sturdy and reassuringly large, but inside is crowded and chaotic, with streams of over-loaded passengers passing backwards and forwards, unconsciously shoving against anyone in their path in their anxiety to find a berth.

It is confusing for everyone. There are a number of different classes of accommodation aboard the *East is Red* number 41, none of which is really comfortable. At worst you might be sleeping on a sweaty floor, competing for a space simply to rest your head. At best, if you have the money and some luck, you will end up in first class (or 'soft' class as it is euphemistically called, for the sake of ideological correctness), with a fan and a mere two beds whose thin mattresses come with straw mats to absorb the sweat. And you will have the use of a small saloon for recreation during the day. In between these two extremes are various sizes of dormitory with wash basins and rickety tin bunks.

After about an hour of continuous pandemonium, there is a brief hiatus of comparative peace, the sun baking everyone into an itchy desperation to be off. On shore, the day has swung into action and a steady stream of traffic is visible on the street above. When the hooter finally blasts the whole ship jumps, except the crew who amble into action in sensibly languorous style. The ropes are gradually fed out, and the *East is Red* drifts out into mid-stream. She takes her time turning around and the engines rumble into life. The astoundingly loud hooter, whose every blast unfailingly causes everyone to shudder with shock, screams at the smaller ships to make way and we begin the journey east.

ABOVE, *a quintessential image of rural China – ploughing the paddy fields with oxen*

BELOW, *the spectacular Xi Ling Gorge, the longest of the Yangtze Gorges*

ABOVE, *a back street in Wuhan: made up of three cities, it is one of China's largest conurbations*

No Colour, No Shadow

Most welcome is the breeze which fans over the ship (although precious little of it makes its way into the ship's inner recesses; for that you need to be travelling 'soft', which entitles you to use a small private deck at the prow of the ship) as we pass through the extensive riverine suburbs of Chongqing. It is a crowded city, one of the largest in China, and where factories are not belching out their fumes houses and apartment blocks jostle down to the water's edge.

But even great cities have their boundaries and finally Chongqing slips away altogether, to be replaced by high river banks of red mud fringed by crops growing right to the edge.

By now the sun was approaching its zenith and the great heat, despite the breeze, was oppressive. There were no shadows, just a constant filmy haze which lent a dreamy aspect to the whole scene; and those flat colours which so characterise China. China is the country of greens, browns and yellows, the shades of the countryside under cultivation. Unlike India and the nations of south-east Asia, this is not a land of vivid colour.

As with all cruises, the focal points of the day tend to be the meals. A small dining room at the stern of the first-class deck fed those who chose to eat something like full-scale meals. Initially the call to eat was greeted with enthusiasm, but the dining room was small and hot, and the food greasy, washed down with warm beer. It seemed churlish to complain, however, as it was unlikely that in the circumstances there was any possibility of improvement. In any case, nobody was travelling on the *East is Red* for the food. If you were not on board simply to reach A from B, you were there for the grandeur of the scenery and for the experience; and also to obtain a glimpse of a largely unchanged China.

The highlight of the trip was to be the Three Gorges, which were a day away yet, but there was plenty to absorb the mind and eye in the meantime: a bright red pagoda, known as the Stone Treasure Stronghold, cascading down the cliffs to our left; the crew of a small sailing boat rowing frantically from beneath our bows as the hooter boomed and boomed, followed by the captain giving the terrified sailors a scolding over a loud hailer as they looked up in mute astonishment; water buffalo cooling off in the shallows at the river edge; and the course of the boat as it shifted and heaved in order to find the safest route through what could be a treacherous waterway. If it were not for the heat it was possible to stand transfixed at the railing all day long.

THREE GORGES

Nonetheless, a stop at a real port was a welcome diversion. Wanxian, which for decades must have seen only a handful of foreigners, was a town where westerners were still the object of unblinking fascination. Here was an opportunity to stock up on fruit and fresh peanuts, to be carried back to the ship in the colourful baskets which appeared to be a speciality of the area. Otherwise, there was the antiques store, or the visit to the silk-weaving factory, where the usual formal introduction, accompanied by tea or Coca-Cola, was followed by a tour of the workshops, places of Victorian machinery making an infernal din and working practices of the same era.

The next morning we awoke in a cloud of mist. Beyond it, apparently, lay the first of the gorges. The air was cool and for once we wanted it to warm up so that we could enter the first of them, the Qutangxia, or Bellows, Gorge. As the sun rose, everybody on board agog with expectation (for the gorges are part of Chinese folklore), the first peaks became visible above the wispy upper fringes of the mist as it gradually evaporated. The engines were rekindled and we entered, like 19th-century explorers, the bowels of the earth.

The Qutangxia is, at a mere 5 miles (8km), the shortest and most magnificent of the gorges. Here the river has narrowed and the waters have quickened. On all sides are massive perpendicular cliffs, their grandeur heightened by the scraps of mist that cling to the peaks, gilded by the sun, while we below are still sunk in stygian gloom. There is a feeling of rushing through, as if the waters are in charge of the boat, not the other way around. It is truly dramatic, a fairground ride with a touch of real adventure. And then we are out. Calm descends, we feel the warmth of the sun again and we cruise on. The day is spent passing in and out of the gorges, for the next, the Wu Xia, is another 19 miles (30km) downstream, whilst the longest, the Xi Ling, is a further 31 miles (50km) beyond that and is itself 48 miles (76km) in length. By the time we are nearing the end of this last gorge, the sun is setting behind us.

We slide along a new Yangtze now, of wide rippleless waters and low banks where the sorghum grows across the flatlands of eastern China. At Wuhan, where the Yellow Crane Pagoda stands, there is a long wait before we resume our journey for the last stage to Nanjing. This is the 'land of fish and rice', millions of acres of muddy paddy fields and villages with their fish pools, the farming El Dorado of China, where the peasant, for centuries the ever yielding backbone of the country, is now becoming rich at last.

We know when we are at Nanjing because of the unmistakable Nanjing River Bridge, a project begun then abandoned by the Russians (who said it was impossible), but which the Chinese took on and managed to complete with, so they claim, barely a life lost. Beyond it is the old Ming wall which still girds the city. Made of fired bricks, each with the signature of its maker, it is an earlier testament to the Chinese genius. Nanjing turns out to be city of leafy charm.

So it is time to leave the *East is Red*, and, although there are a few nostalgic regrets, it will be good to have a bath.

LEFT, *the imposing Yangtze River Bridge at Nanjing, linking Wuchang and Hanyang*

PRACTICAL INFORMATION

■ The possible ways of taking a cruise along the Yangtze are many and varied. The most expensive option is to use one of the first-class tourist ships which ply the river between March and October. These ships have twin-berth air-conditioned cabins, each with shower and WC, a restaurant, a lounge, a gift shop, a clinic, observation areas and a sun deck. Tour operators offer these ships usually as part of a package tour to China. To find out about such tours, or how to obtain individual tickets, contact the China International Travel Service (CITS) which has offices in London, New York and Melbourne.

■ To travel in one of the standard ferry boats, once again contact CITS before leaving home, or contact the CITS office in the river port from which you intend to leave, or from the terminal itself, which is the cheapest option. The type of accommodation on these ships is variable but broadly speaking is divided into four or five classes. Some do now have first-class accommodation with private bathrooms and air-conditioning, but in general the best available is second-class with twin berths, an electric fan and communal showers. Then there is third class with up to 12 beds and fourth class with up to 20 beds. Fifth class may sleep up to 40 people.

■ The journey time from Chongqing to Wuhan (the most popular route) is three nights and two nights.

ABOVE, *Shangai presents an image of unbridled vibrancy born of capitalism*

Where East meets West: Hong Kong to Shanghai

CHRISTOPHER KNOWLES

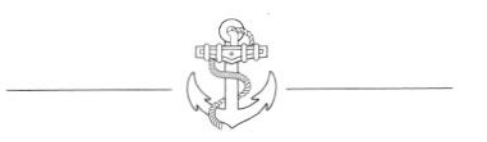

Nearly three days at sea, skirting the south-west coast of mainland China, took us from one great port to another, from Hong Kong to Shanghai. Both cities conjure up strong images, one of skyscrapers and an intense, frenetic vitality, one of illicit excitement, and both bear the stamp of British colonial rule yet remain unmistakably Chinese. Plying between the two are the comfortable and workmanlike ships of the Jing Jiang Line, a Shanghai company rejoicing in the freedom to do business with the outside world since the death of Mao. Having said that, the majority of passengers are Chinese, going about their daily business.

If you manage to stay in a room overlooking the harbour in one of the hotels that line the waterfront, you can see why Hong Kong is one of the world's great ports.

For one thing, it is vast. Harbours are usually finite, an area of quays where ships are loaded and unloaded, where cranes hang portentously, where customs sheds sit back in solemn expectation. A few ships, awaiting their turn, are anchored further out, turning with the tides. Here, in Hong Kong, there seems to be no beginning and no end.

The harbour, one of the largest deep-water ports in the world, consists of a wide, restless strip of water between Hong Kong Island and Kowloon, at the tip of the Chinese mainland. These waters are never still. Yes, there are the liners, cargo ships and naval vessels from Britain and America which slide in and out with majestic precision; but more striking is the incessant movement of the small craft that criss-cross endlessly between them. Wherever you look, there is some craft leaving or entering a berth, whether it be the green Star Ferry on its endless shuttle from one side of the harbour to the other, or a bobbing sampan nosing its way cheekily among craft at

RIGHT, *the green Star Ferry is a familiar sight crossing Hong Kong harbour*

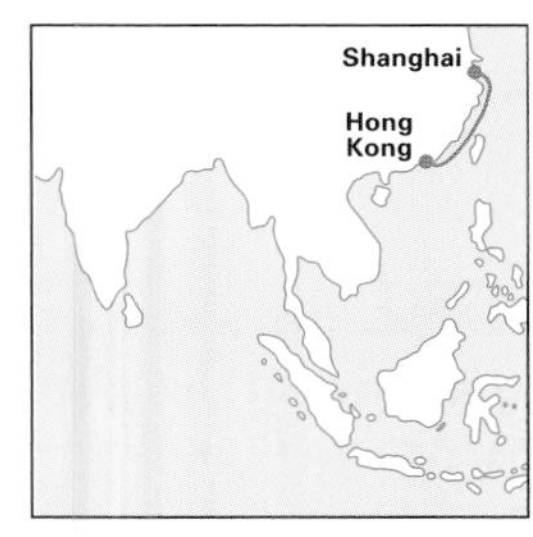

The trip begins in Hong Kong which, between 1842 and 1997, was a British colony. It consists of a group of islands, of which Hong Kong itself, though not the largest, is the most important and by far the most populous. Between Hong Kong Island and the mainland is the teeming strip of water that is Hong Kong harbour. Hong Kong's attraction has long been two-fold – its status as a colonial outpost, and its proximity to China, a place where east and west meet and on the whole complement each other. Hong Kong, more than anything else, is a place to experience. It is very much a modern city, where making money is the object of all who live there. But for the visitor there is plenty to see and do. Shopping and eating are high on the list for many, but there are the local markets to enjoy; there is the Peak Tram to Victoria Peak, the rural calm of the New Territories, and the maritime tranquillity of the Outlying Islands. There are museums and temples, Aberdeen with its fishing boats and the spectacular aquarium at Ocean Park.

The journey to Shanghai follows the south-eastern coast of China, the provinces of Guandong, Fujian and Zhejiang. It passes to the west of Taiwan, threading between the smaller islands of the South China Sea, before reaching the gaping mouth of the Yangtze (Chang Jiang), at 3,938 miles (6,300km) the longest river in China and the third longest in the world.

Almost immediately the ship leaves the Yangtze for an hour-long journey along the short but deep tributary, the Huang Pu, upon which is built Shanghai. Shanghai was never exactly a colony of the foreign powers but was nonetheless given over to them in the mid-19th century to do with pretty well as they wished. The result was a phenomenon, a city, 'the whore of Asia', that grew on the back of unbridled capitalism. The western powers built in their own image, creating a city of largely European architecture from between the late-19th and mid-20th centuries. For that reason alone it is worth strolling along its streets – the Bund, the Nanking Road – although many of these buildings are making way for the new Shanghai, which promises to be as vibrant as the old one. Visit the Peace Hotel with its ancient jazz band; the Old Town with its narrow streets and traditional garden; the new museum and the Jade Buddha temple.

ABOVE, *tourists at Aberdeen, a busy fishing port and harbour on Hong Kong's southern coast*

anchor. But no junks any more, unless you count the cruisers that the expatriates use at weekends to recover from handling all that money.

All Bound for Shanghai

Hong Kong has survived the years pretty well, moving with the times as demanded by the currents of economic fashion. Anyone expecting the Hong Kong of Suzie Wong will be disappointed though – there is nothing old fashioned or sentimental about Hong Kong! Absolutely nothing old survives there, and there is apparently no room for sentiment. It doesn't even seem especially oriental at first sight, more like New York-on-Sea. But in between the skyscrapers there are the markets, gambling joints and a rather frosty glamour showing that at heart Hong Kong is essentially Chinese. The British may have ruled these territories as a colony for over a century but the mark they have left is barely skin deep.

Nowhere was this more evident than at the terminal for boarding the passenger liner to Shanghai – as far as I could see there was no other non-Chinese making the journey. I and several dozen Chinese people stood in line in between a maze of railings with our various piles of luggage, waiting to board the lighter which would take us to the ship. This, run by the Jing Jiang Line, was in good order with comfortable cabins, and a lounge with a library of Chinese books. The purser spoke no English; or at least only in a formulaic way. 'Excuse me, where is the lounge?' 'Ah, welcome aboard, sir!', was the friendly reply. Clearly, non-Chinese passengers were always in the minority.

After an hour we set sail to spend the first of three evenings at sea. You could not have dreamed up a more poetic exit from Hong Kong: the setting sun glowed pinkly across the sky; the steel and glass skyscrapers, stacked for all the world like cigarette lighters up the slopes of Hong Kong island, were molten. The choppy harbour waters glinted like glass and as we sailed among the outlying islands there was a sort of tropical splendour about it all that suggested Hollywood but was a whole lot better for being the real thing.

As darkness descended, the dinner gong sounded. If there is one thing that is important in the life of the Chinese, it is food. Where else in the world is the question 'have you eaten yet?' a form of greeting? We poured into the dining room and

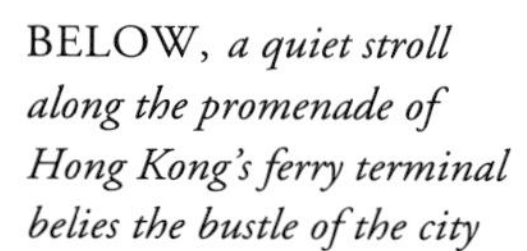

BELOW, *a quiet stroll along the promenade of Hong Kong's ferry terminal belies the bustle of the city*

I took my place at a small table – all the larger, rounder ones were quickly occupied by groups of Chinese who tend to travel in large parties. In the centre of the heavily veneered room was a huge tin vat of steaming rice and another of broth. We were to help ourselves to these, the rest of the meal being served at table.

ABOVE, *traditional fishing boats at Xiamen, in Fujian province, on the south-east coast of China*

FELLOW TRAVELLERS

Having filled my bowls I returned to the table to find that I had been joined by two other passengers. One turned out to be an American, Jim, while the other, a Chinese man, introduced himself as Johnson and turned out to be from Taiwan.

Jim was a gentle man, with a silver moustache and a warm, friendly drawl. He was going to Shanghai to be reunited with someone whom he had met long ago. He was that charmingly relaxed breed of American, betraying no sign of excitement at the journey he was undertaking, or the fact that this was to be his first visit to far Cathay, yet taking a penetrating interest in everything that crossed his path. He was a man with few preconceptions, looking at everything with a fresh eye.

Johnson, on the other hand, was an enigma. Talking to him was, for a person like myself who *did* have preconceptions, like talking to a Chinese character from an old comic book. Though not yet anywhere near middle-aged, he had the erect posture of a mandarin – he was not quite impassive, but certainly he was inscrutable. He had left Taiwan behind, he said, left it behind for good. Curious at such a quaint idea (after all, a less surprising encounter might be with someone leaving China for Taiwan and the luxuriant chimera of capitalism), we pressed him on the matter. The truth, he continued, was that he wanted to be buried in the motherland. He was apparently serious, but Jim and I burst out laughing simultaneously. The Chinese traditionally take their place of burial seriously – indeed for the emperors it was practically their main consideration – but given Johnson's age, about 35, 40 at the most, and seeming rude health, such concerns were premature to say the least.

Johnson gave a snigger, looking about him with a look of flickering furtiveness. He insisted

ABOVE, *the paper-like sails of this fishing barge have been ravaged by time and the elements*

INSET, *Shanghai's Peace Hotel (formerly the Cathay), was once host to Noel Coward*

that this was the truth. He got up to help himself to more rice. 'That guy is an operator', drawled Jim, with a twinkle in his eye.

After dinner I strolled on deck. The air was cool, and the sea calm. We were not far from land – I could see the lights of China away to the left. Young couples huddled together on deck chairs, whispering and laughing.

Our days were long, divided by meal times. On sea voyages time seems to stand still, just the breeze, the sparkling sea and the constant judder of the ship. From time to time we could see land, either the mainland to the left, or the occasional island drifting away to the right. Sometimes there were flotillas of fishing boats, not far away, unwieldy junks with their papery sails, dabbling like birds in the water.

I looked forward to lunch and dinner (breakfast was a free for all, so we usually did not meet), just to listen to Johnson's supply of stories about his life, most of which made Jim laugh uproariously, because they were so delightfully preposterous. Johnson would put his arm around Jim's shoulders and ask 'Oh Mr James, why you not believe me?' But in between his fables, there were glimpses of the truth. He had done his national service in the Taiwanese army, then he had been a merchant seaman and visited Liverpool and New York. We also learned that he had relatives awaiting him in Shanghai who were going to help him become reacquainted with the home of his ancestors.

Between meals, I would try to find a patch of sunlight in which to sit and read. Whilst the ship ploughed on through the China Sea, creating its own laws of time and motion, my fellow passengers were never still. Not that they did much – there was not, after all, much to do – the swimming pool and tennis court being closed this early in the spring – but it took this journey by ship to make me realise what a restless people the Chinese are. In pairs or small groups, they were forever ambling about the deck, eyes continually darting from object to object. There was an endless stream of chatter and laughter and there was always someone draped over the sofas and deep armchairs, legs a-dangle. Everyone seemed blissfully content and yet, at the same time, profoundly impatient.

MEETING THE YANGTZE

On the last afternoon we arrived at the meeting point of the China Sea and the muddy effluent of the mighty Yangtze (Chang Jiang), as distinct as two layers of chocolate. Before long the engines stopped for the first time since leaving Hong Kong and we anchored for the night in the gaping mouth of China's greatest river.

Early next morning, in bright sunlight, we steamed upriver, just as the tea clippers had done a century before, to meet the Huang Pu, the tributary feeding Shanghai. For an hour we thrummed between ageing rust-buckets, spanking new liners from all over the world, decrepit submarines, ferries and chains of barges, filled with who knows what and sunk impossibly low in the water.

Then we rounded a bend and there was Shanghai, just as it would have appeared 50 years before when it was ruled by everyone except the Chinese. It was an incongruous sight, in Red China, to see a waterfront that looked for all the world like a sun-drenched Liverpool. But it was a rather stirring sight, this line of icons to plutocracy – there was the old Cathay Hotel, where English playwright Noel Coward wrote *Private Lives*; next door to it and exactly the same height – or perhaps just taller – the Bank of China, built in the image of the Empire State Building by the banker H H Kung, who was determined not to be outdone by Sir Victor Sassoon, owner of the Cathay. Beyond them the domed roof of the former Hong Kong and Shanghai Banking Corporation premises and the Custom's House, surmounted by 'Big Ching', which still plays the Westminster chimes. At the far end, close to the old French Concession, is the building that once housed the Shanghai Club, boasting the longest bar in the world.

We docked and disembarked, then went our separate ways. Well, sort of. I saw Jim a couple of times in Shanghai – he made contact with his friend and was able to arrange for him to go over to the States for a visit. Of Johnson I saw a great deal more, since he insisted that I go to meet his relatives. They were already wary of their eccentric cousin from Taiwan, so perhaps they were not too surprised that he should arrive with a westerner in tow. Anyway, they very kindly placed me in a hostel run by a local Christian church. I grew to know and like Johnson's relatives, from whom I was able to discover the real reason for their cousin's 'homecoming'. But that, as they say, is another story, a sequel to an encounter on a voyage to Shanghai.

LEFT, *the Huang Pu tributary of the vast Yangtze meets the sea at Shanghai*

PRACTICAL INFORMATION

- The journey between Hong Kong and Shanghai can be accomplished in either direction. There is currently a departure from Hong Kong every five days, with some of the boats stopping at Ningbo en route. Departures from Shanghai leave every seven days.

- There are three ships on this route at the present time. All are reasonably well appointed, the cabins being clean and comfortable, with their own bathrooms. Each boat has a lounge, a bar and a swimming pool. The cost of all meals is included in the basic price and the standard of food is fairly good. There are up to eight classes of cabin. If tickets are purchased in Shanghai the prices are variable because sometimes you will be charged a special foreigners' price, but sometimes not.

- The journey takes two-and-a-half days. The best time to travel is in the late summer after the rains and before the autumn, which can be very cool in Shanghai.

- Tickets are available from CITS in Shanghai (also London, New York and Melbourne) or from the China Ocean Shipping Agency at 13 Zhongshan Dongyi Lu (Tel: 329 0088). In Hong Kong, contact the China Travel Service (CTS) at CTS House, 78–83 Connaught Road, Central, Hong Kong. Tel: 853 3533.

A Journey Back in Time along the Mekong River

BEN DAVIES

ABOVE, *gold is used in abundance to decorate the extraordinarily beautiful temples of south-east Asia*

BELOW, *traders on the Mekong at Laos; the river is a major transport artery*

From high up in the Tibetan Himalayas, the Mekong River plunges down through China, Myanmar (Burma), Thailand, Laos, Cambodia (Kampuchea) and Vietnam, passing through some of the least known areas of south-east Asia on its 3,000-mile (4,800km) journey to the South China Sea. As you travel from Chiang Saen to the Mekong Delta by speedboat, cargo boat and sometimes by bus, you will witness some of the most magnificent stretches of the river, as well as a mind-boggling collection of ethnic peoples, temples and cultures. Although this region is lurching rapidly into the modern world, this river still evokes images of a bygone age.

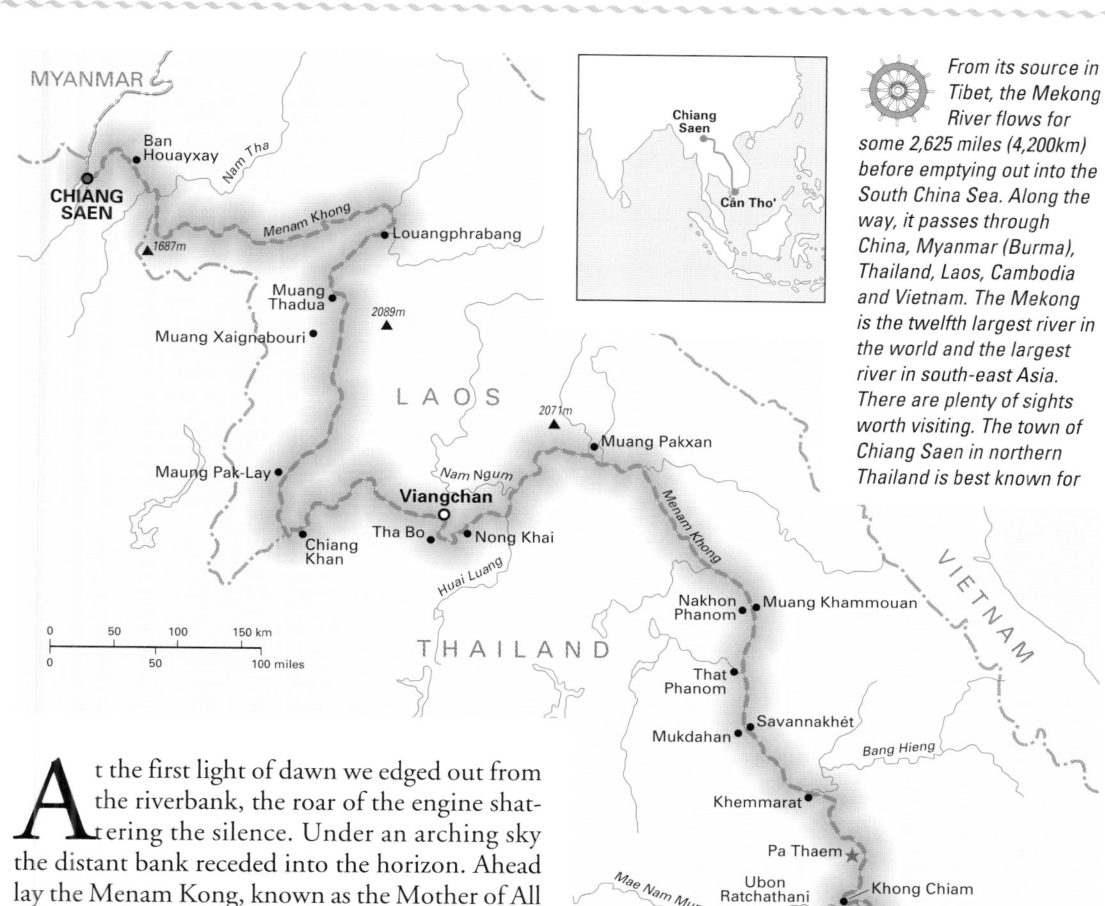

From its source in Tibet, the Mekong River flows for some 2,625 miles (4,200km) before emptying out into the South China Sea. Along the way, it passes through China, Myanmar (Burma), Thailand, Laos, Cambodia and Vietnam. The Mekong is the twelfth largest river in the world and the largest river in south-east Asia. There are plenty of sights worth visiting. The town of Chiang Saen in northern Thailand is best known for its temples (wats). Those to see include Wat Phra That Chom Kitti, which is reputed to hold part of the Buddha's cheekbone, the 13th-century stuccoed chedi *(pagoda) of Wat Pa Sak and the imposing 14th-century Wat Chedi Luang.*
At Luang Prabang, the former royal capital of Laos, wander around the fine temples of Wat Xiengthong and Wat May, as well as Wat Siphuthabaht. Other sights not to miss include the old royal palace and the 400-year-old cave temple of Tham Ting.
Chiang Khan and Nong Khai have some fine old river houses as well as scenic routes in the vicinity.
In Kong Chiam the different coloured waters of the Mekong mix with the River Mun. Other places of interest are the Pha Taem prehistoric paintings and the Kaeng Tana National Park.
In Phnom Penh, capital of Cambodia, visit the royal palace, the national museum of arts as well as the shocking museum at Tuol Sleng Museum, witness to the genocide of the Khmer Rouge.
In the Mekong Delta of Vietnam, sights to see include the floating market at Phung Hiep and the picturesque canals that criss-cross the entire region.

At the first light of dawn we edged out from the riverbank, the roar of the engine shattering the silence. Under an arching sky the distant bank receded into the horizon. Ahead lay the Menam Kong, known as the Mother of All Rivers, lazily cutting its way through a countryside of wooded hills and rice fields.

Of all the rivers in south-east Asia, the Mekong is the most legendary. It springs to life high up on the Tibetan plateau where the Himalayas are at their most beautiful. Still in its youth, it passes through the magnificent gorges of Yunnan Province in south-west China then flows alongside the border with Burma, threading its way through a countryside dotted with ethnic tribes. At the crossroads of Laos and Thailand its course widens, its gentle current mocking the ferocity of earlier rapids.

I had started out in Chiang Saen, in the north of Thailand, where the lapping of the river mixes with the thunderous roar of speed boats. In this town of ancient temples and pariah dogs I joined a group of Thai students and a tour guide for the first leg of my Mekong journey.

The Golden Triangle

To the sounds reminiscent of a jumbo-jet at take-off and the vibrations of an ancient spin-drier, we sped off downriver propelled by a giant 10ft (3m) prop shaft connected to a Toyota car engine. Lurching past Wat Pa Nga and the famous Wat Phra That Chom Kitti, a temple believed to house a part of the Buddha's cheekbone, we promptly disappeared in the mist. Since the mid 1960s, this infamous area, nicknamed the Golden Triangle

BELOW, *for those living along the Mekong, much of daily working life is spent in a boat*

ABOVE, *formerly a royal building, Wat May, at Luang Prabang, took over 70 years to complete*

has, according to some estimates, regularly supplied a staggering 2,000 tons of opium annually to the world's addicts, generating more funds than even rice or tourism. Neighbouring Burma is cited as the chief culprit in the narcotics trade but the accommodating Thai people have shown themselves by no means unwilling to make a little money from it.

My fellow passengers, however, do not resemble drug smugglers, nor even undercover agents. One student claims to have fled the pariah regime in Burma, another comes from a small village a half-day's journey downstream. Soon they are picnicking on chillied chicken's feet and potent Sangthep whisky (the staple local beverage) with all the enthusiasm of toddlers at a tea party.

It was late morning when we crossed the border of Laos at the town of Ban Houei Xai. Beneath a fat palm tree, a group of backpackers sat incongruously drinking cans of Coca-Cola through pink straws. All around them villagers pottered around on antiquated bicycles, seemingly blissfully ignorant of the tourist invasion that had suddenly befallen them.

If the first thing that you notice about Laos is the undeveloped nature of the country, the second is the currency. The *kip*, as it is known, is worth about as little as the paper it is printed on. To simplify matters, however, the understanding Laotian people allow foreigners to pay in dollars or even in Thai *baht*. Often, at the end of your journey, you will end up paying in all three currencies. Alternatively, you may end up receiving your change in the form of boiled sweets.

There are other delightful inconsistencies in this isolated backwater. When a boatman says he will leave in a little time from now, he means tomorrow or the next day, or if there is a temple fair, maybe the day after. But as soon as you think you have mastered this wonderfully flexible attitude to time, you arrive at the anticipated moment to find that the boatman and the only form of transport have long since departed.

Thankfully there is more than one boat leaving for Luang Prabang and so I set off once again downriver. Soon Thailand is little more than a dot in the distance. As we motor downstream, the countryside becomes more mountainous. Gigantic limestone cliffs, clothed in lush vegetation that spills down to the water's edge, tower above the river.

In Pak Beng, a small village built into the mountainside, we halt for a plate of spicy pork intestines and a glass of Laotian tea. Then we continue on through the wind and gathering rain. Finally, after what seems an eternity, the river broadens out and we draw alongside the landing ramp at Luang Prabang, the former capital of Lang Xiang, kingdom of a million elephants.

CITY OF MAGIC AND PLEASURE

There can be few more magical cities than Luang Prabang, sitting at the confluence of the Mekong and the Nam Khan rivers with its glitttering temples. The Lhao have long viewed the place as the cradle of their civilisation and over the centuries have built dozens of resplendent temples and palaces there. Some, like Wat Chom Si, are gloriously set atop nearby hills surrounded by swaying palm trees; others nestle amidst country lanes where the local people float past on bicycles.

While the people of Chiang Saen lay claim to the Buddha's cheekbone, the people of Luang Prabang boast his footprint. At the temple of Wat Siphuthabaht, perched nearby on a rocky outcrop in the centre of town, devotees flock to the 10ft (3m) imprint which, legend has it, the Buddha left as a sign of his immortality.

Elsewhere in this magnificent city, where the monks file at dawn along the narrow streets in robes of unimaginable glory, dreams take on an uncanny reality. Here, surrounded by perfumed forests, the days of the week, even the months, seem to merge. The lethargy is as immediate as it is all-embracing. Progress seems a world away, hidden by the mists of time.

As striking as the temples of Luang Prabang are the people. Tall and gracious, their fine Mongolian features and rounded faces set them aside from the neighbouring Thais. Despite being gentle, tolerant Buddhists, they are by no means averse to life's pleasures. Until recently, opium dens were as prolific as houses of worship and Lhao women fondly remembered as the prize of

ABOVE, *temporarily aground at Chiang Khan, in Thailand*

BELOW, *Nakhon Phanom, home to thousands of Vietnamese refugees in the 1950s and 60s*

ABOVE, *the Laotian-style* chedi *at That Phanom; the temple has been restored many times*

south-east Asia. 'The extreme liberty of morals that reign here make foreigners easily find hospitality,' wrote Marthe Bassenne in 1912. 'Since libations and ritual gifts appease them, the young girls are seldom shy.'

Whether for lack of ritual gifts or as a result of changing times, my experiences are of a less forthcoming nature. And so after exploring the temples, the sleepy markets and shop houses where they sell potent rice wine, I head back down to the river, back to where the Mekong takes on a new guise as it continues downstream towards Chiang Khan, Sangkhom and Nakhon Phanom.

Early Explorers

In the cool of the morning we set off once again, sailing past ragged villages where plumes of smoke rise into the air from primitive stoves and where the cries of children can be heard above the steady roar of the engine. With each curve, the golden spires of Luang Prabang fade into the distance, until soon they are only memories.

Indeed it is one of the great ironies of the Mekong that a river of such magnitude should for so long have remained shrouded in mystery. In 1866, a naval expedition under Captain Doudart de Lagree set out from Saigon to explore the Me Kawng which, it was believed, would lead them into the heart of Yunnan. Despite their valiant efforts, Captain Doudart and his crew were forced to abandon their boats. Undaunted by this they reached Chieng Mung in dug-outs within six months. From there they continued to Yunnan Fu and Tali Fu, although Captain Lagree himself died before completing the journey.

Later expeditions foundered on the Keng Luang rapids and further downstream near Khemmarat. Indeed, it was only in April 1995 that a Franco-British expedition announced that they had discovered the source of the Mekong, 16,447ft (5,000m) high at the head of the Rup-Sa Pass in Tibet.

My voyage suffers from fewer mishaps. After crossing over from Pak Lay in Laos to Chiang Khan back in Thailand, I continue east along the banks of this great river through a countryside of staggering beauty. Travelling sometimes by boat, sometimes by bus, I stayed in various 'guehow' (the Thais can't pronounce guest houses) overlooking the Mekong. From simple riverine huts I witnessed sunrises and sunsets with compositions and colours so awesome that no artist could have dreamt them up.

The enchanting people who live along this stretch of the river are known as Isaan people and are supremely superstitious. Typically they name their children 'mouse', 'pig' or 'crab' to ward off the evil spirits. They also believe that, unless someone drowns in the Mekong every year, the rains will not come. In November, they float boats made from banana leaves on the river to cleanse their souls for the coming year. On other occasions they place carved wooden phalluses along the river bank to fertilise the water and ensure a plentiful catch.

Below the towns of That Phanom and Khemmarat the river changes course, weaving its way past small islands covered in low bamboo scrub and sparse vegetation. Shortly before it exits Thailand, the sluggish brown waters of the Mekong are joined by the clear waters of the River Mun, Thailand's largest tributary. For a brief moment, the water is clouded as the two powerful currents come together. Then the majestic Mekong continues its course, finally swinging out of Thailand near the town of Kong Chiam and continuing its long journey over the Khone rapids and into Cambodia.

KILLING FIELDS

The sun was bleached out with a heavy haze. It hung over the river, a torpid weight suspended on a cloud of thin air. Mosquitoes mustered around the sprawling port where a few overcrowded passenger boats unloaded baskets of fish and fruits. Here in Phnom Penh, where the waters of the great Tonle Sap River feed into the Mekong, the river brings a sense of melancholy. For this was the scene of one of the world's modern holocausts. Between 1975 and 1979, more than one million Cambodian people died at the hands of the Khmer Rouge. Hundreds of thousands of others were

BELOW, *passenger boats at Phnom Penh, Cambodia*

ABOVE, *Can Tho, on the great Mekong Delta*

sent to re-education camps. Even now the tragedy continues as mines unleash their fury on innocent victims in the rice fields.

In Cambodia more than anywhere the stately course of the river mocks the killing that has taken place along its fertile banks. Now a vast and gentle life force, the majestic river flows for about 313 miles (500km), passing though Kratie, Stung Treng, Kompong Cham and Phnom Penh on its long course towards the Mekong Delta. When I came here three years ago, I could hear the sounds of shelling at night. Today the guns are silent, but the ghosts live on, a haunting silence that accompanies the gentle ebb and flow of the Mekong on the last leg of its journey south.

Crossing into Vietnam, I continue my voyage south through a bountiful land of rice fields and lush plantations that crowd down to the water's edge. Dozens of boats busily criss-cross the great network of canals, carrying fruits and vegetables to the villages that have sprung up all along the low banks of the delta.

The Vietnamese call this stretch of the river *Cuu long*, meaning nine dragons. Each 'dragon' represents one of the major branches of the Mekong as it fans out towards the sea. During the Vietnam war this was the scene of some of the fiercest bombing. Now the soil is once again rich with the silt of the Mekong, the area supporting nearly one-quarter of the country's population.

My voyage down this last stretch of the great Mekong is like a journey into the past. Catching a passenger boat in Chau Doc, I follow the Bassac River, which is known as the lower Mekong, past villages that appear to have stepped back in time. On the river banks, old men in trilbies chase ducks down to the water's edge, whilst at low tide children search for cockles and crabs in the silt to sell in the local market place. A few ancient trucks piled high with crates of Coca-Cola are the only signs that Vietnam too is hurtling forward to join the modern world.

South of Can Tho, the landscape becomes even lusher, the land surrounded on all sides by water and marshland. And so I end up passing Tra On and beyond to where this majestic river empties out into the South China Sea, a palpable lifeline that has silently endured wars and massacres and which now brings hope to the people who live along its banks.

Practical Information

- Long-tailed speed boats from Chiang Saen to Chiang Kong take approximately two hours. From Chiang Kong, it takes five hours to Luang Prabang and a further five hours to Chiang Khan and Vientiane. Slow boats take three to four days to cover the same route.
- From Kratie in Cambodia, passenger boats run to Phnom Penh, taking one day, from where you can continue over the Vietnamese border by bus.
- Slow boats leave from Chau Doc to Hathien and Vinhlong. To cover other stretches of the river you must go by road, travel by cargo boat or arrange your own transport.
- The best time to travel is between October and January when the rainy season is finished – and the river level is high. In the dry months, much of the river is unnavigable.
- Visitors to Thailand, Laos, Cambodia and Vietnam should contact the relevant embassies about visa requirements. You will need anti-malaria tablets and other inoculations. Ask your doctor for details.
- For information about trips, contact Symbiosis Expedition Planning, 113 Bolingbroke Grove, London SW11 1DA (Tel: 0171 924 5906; Fax: 0171 924 5907).

LEFT, *tranquillity and solitude on the middle Mekong*

Discovering Paradise: Indonesia's Secret Islands

FIONA DUNLOP

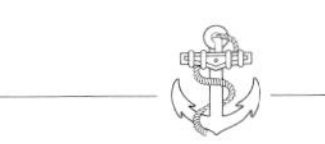

ABOVE, *a single make-shift sail and a paddle – all that is needed for getting from island to island*

With some 17,000 islands scattered around its vast waters, it is no surprise to learn that Indonesia has some truly well-kept secrets. One of its many treasures, caught in the embrace of two arms of octopus-like Sulawesi, is a cluster of islands and myriad atolls called the Togian Islands (Kepulauan Togian). Here in the coral-rich lagoons live Bajau fishermen, commuting between their main island homes and their 'weekend' fishing-shacks. A boat tour of this archipelago offers a truly exceptional insight into their simple lives and into some of the world's most beautiful, still undiscovered coral reefs. But it is a journey not without its discomforts.

Often, the more obscure and enticing the destination, the more difficult access is, and the Togian Islands are no exception. Floating in the equatorial deep blue of Tomini Bay in the northern curve of Sulawesi, without phones let alone planes, they are connected to the rest of the world only by a twice-weekly ferry. Few foreigners visit them, and even fewer guidebooks mention them. However, while staying in Sulawesi's tourist-hub of Toraja-land, my fellow traveller and I had the luck to encounter Rudy, who turned out to be 'the man with the boat' and our key to this undiscovered paradise.

BELOW, *white sand, palms, driftwood, crystal-clear water ... what more could you ask?*

OBSTACLES OVERCOME

A native of Poso, the mainland launching-pad for the Togian Islands, Rudy has been exploring the region, its inhabitants and its coral reefs since childhood and now works as a tourist guide, only recently extending his activities to boat trips. Long-haired, slim-limbed and softly-spoken, looking more Amazonian than Indonesian, he is the antithesis of the average hustling guide operating around the tourist pulls of southern Sulawesi. Despite the off-putting and persistent drumming of seasonal rain, his descriptions of the islands and semi-mystical approach soon convinced us that they were a must.

The next day we were off, with a driver, to cover the 375-odd miles (600km) to reach Rudy's boat. This sounds an unexceptional distance, but the common geographical currency of rugged

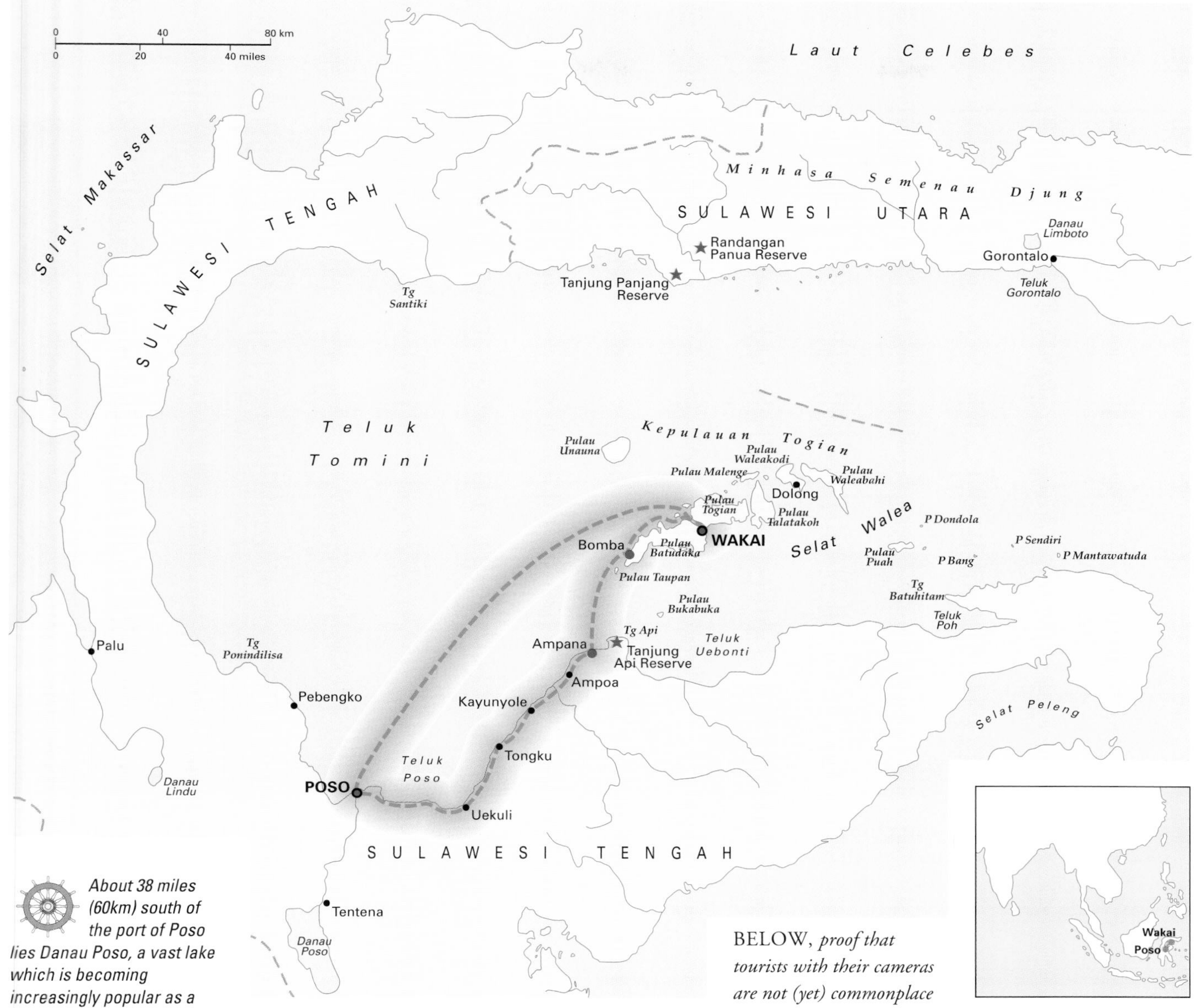

About 38 miles (60km) south of the port of Poso lies Danau Poso, a vast lake which is becoming increasingly popular as a recreation ground for fishing, trekking and swimming. Ancient burial grounds, an orchid park and a 691-ft (210m) long roofed wooden bridge sheltering eel-traps at Tentena are alternative offerings. About 13 miles (20km) east of Ampana lies the 10,484-acre (4,246 hectare) nature reserve of Tanjung Api ('fire cape'). Home to numerous Sulawesi endemic species (black macaques, tarsiers, anoa, a dwarf buffalo and babi rusa, a hairless wild boar sporting four tusks) alongside pythons and deer, it is named after a series of jets of natural gas which catch alight as they come into contact with the air. An easy trail follows the coast through rich forest with occasional views north to the outlying Togian Islands. One of the first foreigners to take a close interest in Sulawesi was the English naturalist Alfred R Wallace (1823–1913), a contemporary of Charles Darwin. He postulated an imaginary ecological division running down the Selat Makassar (Malakassar Strait), subsequently dubbed the 'Wallace Line', that divided the flora and fauna of Asia and Australasia. Sulawesi was believed to have been isolated from the two great land masses of the last Ice Age, so developing specific and unique forms of flora and fauna. This theory was later superseded.

BELOW, *proof that tourists with their cameras are not (yet) commonplace*

ABOVE, *passengers – with a good deal of their wordly goods – awaiting one of the ferries that regularly ply between the islands*

mountains, rift valleys, primary rainforest and lakes, combined with torrential downpours and atrocious roads, rapidly become major obstacles. Within an hour of setting off our route was blocked by a landslide. A mere 250-mile (400km) detour followed with our driver's foot hard on the pedal while he made rapid changes of his two cassettes to calm our (or his?) frazzled nerves. Six hours of tough road later we stopped for sustenance, and bought him two more cassettes.

Our arrival at the tiny port of Ampana seemed like a miracle, the previous night having been spent fitfully snoozing in the van, our way blocked by yet another landslide. In this case no terrestrial detour offered itself as a solution: there was only one road from Poso to Ampana. So, after bidding farewell to our valiant driver, now a true comrade in arms, we clambered over the wall of trees, branches, rocks and squelching mud and squeezed into a local bus. This proved to be a definite step down in terms of comfort, with an in-house rooster pecking at my companion's ankles and no cassettes. However, after surviving the obligatory breakdown, we at last creaked into Ampana, where Rudy set about looking for his boat and buying provisions for our three-day forage.

'Is this really worth it?' we mused, falling out of the pony-trap that had taxied us along what seemed to be an exceptionally rutted track to the beach. Rudy, unsuccessful in locating his boat and its crew, assured us that we would catch up with it on the nearest of the Togian Islands. In the meantime we were being fobbed off with a crossing in a wooden fishing boat doubling as a ferry and loaded with passengers, baggage, jerrycans and sacks of produce. Off we went, our cameras and bags held high above the spray by the amiable boatmen, eventually to collapse into a deep sleep in an attempt to recover from the traumas of the journey behind us.

A World Away

When I awoke a couple of hours later we were drawing into the harbour of Bomba on Pulau Batudaka, the main island of the Togian archipelago. The late-afternoon skies had optimistically broken their leaden mould to reveal patches of blue, while outriggers and fishing-boats bobbed in the harbour and coconut palms swayed in the background. On the beach fronting the straggling fishing village the customary hordes of bellowing Indonesian children jigged around with joy. 'Hullo Mister!!!' they screamed in hysterical disunison, oblivious to my gender, just delighted to see Western faces in this far-flung outpost.

The local inhabitants are predominantly Muslim and many are Bajau, an ethnic group of former sea-gypsies that is scattered around the Celebes and Sulu Seas from Sulawesi to Borneo and the Philippines. Although their origins remain obscure, their lives and beliefs are always intricately related to the sea: many never even set foot on land. In Bomba, they exist on producing palm-sugar and desiccated coconut and catching fish. Elsewhere in the Togians they sell terrapins and sharks' fins to the ubiquitous Chinese middlemen who rake in the profits from customers in Jakarta or Hong Kong.

Our first night in the Togians was to be spent in a Bajau homestay. As we entered the relatively well-kept house and were shown into the main bedroom relegated to us for the night, strains of French conversation emerged from the back room, followed by the familiar 'plop ... plop' sounds of a lobbed tennis-ball. It was a television, and the family was glued to Roland Garros for the French Open, transmitted by satellite to this tiny corner of the earth. This was not the last anomaly that we would encounter, but it was the most unexpected from a cultural point of view.

Rudy busied himself around the harbour, having at long last tracked down his errant crew, and after a simple family dinner we slid into dreams of choppy seas and crashing trees, wondering whether our efforts so far would pay off.

When light dawned the next morning, the sky was clear and the waves were scintillating. Rudy, his skipper and boathand were at the ready and a large mattress piled high with cushions and a tarpaulin were installed on the deck in front of the wheel-house of his converted wooden fishing boat. Stretched out in such unexpected comfort, we felt like pashas. As we putt-putted northwards out of Bomba's harbour, beckoned by turquoise seas studded with tiny uninhabited islands, our morale soared. Soon an unmistakable conical silhouette rose on the horizon. This was Pulau Unauna, a volcano that last erupted in 1983. Now, crowned by a gentle halo of cloud, it looked reassuringly calm. A few miles beyond was the invisible line of the Equator.

More sharply sculpted atolls came into view, their limestone eroded over the millennia by the sea to create jagged mushrooms crowned with tufts of bushy vegetation. Larger islands were studded with pockets of glistening white sand. Like huge ships adrift, they floated around us. Then, all of a sudden, the sea began to churn and seemingly out of nowhere came a shoal of dolphins, skipping, ducking and diving beside us through the aquamarine tints. Rudy told us that this vast bay was also home to turtles, dugongs (sea-lions), sharks and whales. Our appetite for the Togians sharpened by the minute.

ABOVE, *signs of 'industry' – platform-building in the shallow waters*

That day's destination was a group of shacks erected in shallow waters beside a coral-reef. These rickety bamboo platforms topped by thatched palm-leaf walls and roofs dot the Celebes Sea from north to south and in many cases are the only homes of their Bajau builders and owners. Some farm seaweed, others dive for shells, but the majority practise intense fishing, their catch laid out to dry in the sun before being sold to the mainland. In the case of the Togians, the shacks are used as occasional homes, the entire family sailing to them in outriggers billowing with patched sails. During our trip we were to pass many convoys, the women often in conical hats shading ghostly white faces produced by rice-paste sun-screens, the children packed into the helm.

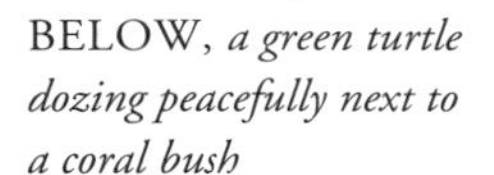

BELOW, *a green turtle dozing peacefully next to a coral bush*

ABOVE, *the quiet waterfront at Wakai, belying its role as the commercial centre of the Togian Islands*

INSET, *the Togian Islands Hotel, built in the harbour of Wakai*

After mooring beside our designated 'coral-house', as Rudy poetically named the precarious construction, we were pulled up a shaky ladder to the no less unstable shelter. Our communication lifeline throughout was Rudy, the only person to speak anything other than *bahasa* (Indonesian), although at this stage in my long Indonesian journey I was able to utter a few stock phrases. Rudy's captain, Ahmat, with his sharply chiselled face and eyes that seemed bleached by the horizon, and boathand Toto, muscular and eternally beaming, were among the ten people who settled down on the mats that night. Dinner had been part fresh fish and part less savoury sago, the Bajaus' staple diet. Although easily farmed and cheap, it is low on vitamins and since few other vegetables are eaten, many Bajau children have curiously streaked blonde hair, a sign of vitamin deficiency.

Underwater Hazards

On arrival that afternoon I had made my first dive into the sea from a canoe paddled by our fisherman host, out to catch our dinner. Transparent shallow waters washed over a mainly grey and dead reef, destroyed by years of dynamite fishing that has now become strictly illegal and is even policed internally by the Bajau themselves. As I snorkelled along, revelling in the sight of brilliantly coloured fish flitting beneath me, a disturbingly familiar silhouette appeared. A shark. Those fins. I was sure of it. Back I swam, a lot faster than usual, and eventually clambered up the ladder to safety. Typically, Rudy was unperturbed: 'It was probably just a reef shark. Don't worry. They don't attack.'

His attitude changed somewhat, though, next day, as we were snorkelling at our next stop, an island beach edged by the most beautiful coral reef I have seen in my life. Gliding blissfully through this underwater fantasia of giant platters, mushrooms, cacti, branches, tubes and soft palpitating sea anemones of brilliant purples and oranges inhabited by luminously coloured fish, Rudy suddenly grabbed my arm and hauled me towards a rock. 'Did you see it?' he gasped, his habitual calm evaporated. 'What?' 'The shark?' Luckily I hadn't, but Rudy's heightened perception put an end to that particular watery interlude. Barbecued fish and rice cooked up on the powdery white sand by Ahmat and Toto soon reconciled us to this corner of paradise before we continued our journey safely ensconced in the cushions of the boat. We saw no whales or dugongs, but the narrow channels between islands gave us closer glimpses of the dense vegetation clinging to the atolls.

Back to Reality

By late afternoon we chugged into the busy port of Wakai, a relatively prosperous little town which controlled most Togian commerce and also claimed the archipelago's only hotel, a pristine white clapboard affair built on stilts in the harbour and featuring wooden walkways. Out strode a smartly dressed woman, Huntje, who was to be our hostess for the next two nights before we caught the ferry back to Poso. A half-Chinese and half-Indonesian Christian, Huntje was an extraordinary character, particularly in the evening when she got stuck into her deserted karaoke bar and belted out rock and roll numbers between sporadic electricity cuts. Our status as the only hotel guests gave us her undivided attention and we were taken to her family house where her Chinese husband introduced us to the art of raising coconut crabs, a lucrative gastronomic speciality much appreciated by the gourmets of Jakarta. Dynamism was the couple's common trait, and with a couple of private islands tucked under their belt, scuba-diving gear being organised and a speedboat ready to ferry visitors from Ampana, it seemed that Wakai's touristic future was assured.

Our equatorial island idyll was gradually winding down but the grand finale came in the form of the return trip to Poso. As we headed towards the wharf where the battered ferry had moored, scores of bag-laden people engulfed us. On board, somewhat aghast, we surveyed the one and only deck. Mats, baggage, sleeping bodies and smiling faces blanketed every available space. Rudy, still with us, muttered something about private cabins. Sure enough, magically, a low-key form of ship's purser materialised and for a few rolls of *rupiah* notes a door was unlocked. Watched over by a poster of Jesus Christ, the cabin offered us four dubious-looking bunks with unwashed sheets. We had no option, and out came our trusty and ever-versatile sarongs, to be used as bed linen and screens.

Our last view of the Togians was from the prow of the small ship. Lapping up the salty breeze and fast-fading light, we vied in high experience and tall tales with the only other Westerners on board, a young German couple who had been staying in a *losmen* (budget guesthouse) at Katupat on the neighbouring island. Again the sea churned and again a shoal of dolphins skipped along beside us, a fitting farewell from this obscure hideaway of untainted lives lost in the mesmerising kaleidoscope of Indonesia's thousands of islands.

Practical Information

- Public passenger ferries make twice-weekly runs from Poso to Gorontalo, stopping at Ampana, Wakai, Katupat and Dolong (the last three in the Togian Islands). Departures from Poso are on Mondays and Thursdays at 10pm and from Gorontalo on Monday and Friday evenings.
- The nearest, most frequently served airport is at the diving-centre of Manado (about 228 miles/365km from Gorontalo), although Poso and Palu have weekly flight connections with Ujung Pandang.
- Indonesian Tourism Aid Promotion Office, Elizabeth Street, Sydney, NSW. Tel: 2 9233 3630.
- The tourist guide Rudy Ruus can be contacted by letter at Jalan Pulau Seram, Nusa Indah 8, Poso, Sulawesi Tengah, Indonesia.
- Wakai's one and only hotel, the Togian Islands Hotel, can be written to at Wakai, Kec. Una Una, Kab. Poso, Sulawesi Tengah, Indonesia.
- Poso offers reasonable accommodation at the Hotel Bambu Jaya, Jalan Agussalim 65 (Tel: 452 21570). The Poso tourist office at Jalan Kalimantan 15 (Tel: 62 452 21211) is helpful for other information on this richly rewarding region.
- The driest and therefore best periods to visit this part of Sulawesi are September to November and February to March.

LEFT, *passenger ferries often seem loaded with everything except the kitchen sink*

ABOVE, *traditional clothing, Balinese style*

Chasing Dragons: from Bali to Sydney

GARY BUCHANAN

Starting, appropriately, from Bali, the 'island of the Gods', the supremely sophisticated and luxurious *Silver Cloud* cruises eastwards to Komodo Island, home of a fearsome dragon, a unique prehistoric lizard. On then to Darwin, capital of Australia's Northern Territory. Here passengers can take a trip into the Outback to Kakadu National Park, a wonderful land of Dreamtime and fantastic wildlife. Continuing on down past the Great Barrier Reef, the ship docks at Cairns, known as Queensland's Playground, then Brisbane, its capital, before finally ending the cruise on a very up-beat note at Sydney.

BELOW, *immaculate terraced rice fields form part of Bali's landscape*

There is a legend about an island, east of Java, 8,000 miles (12,800km) from Britain. It was a very beautiful island, but its fertile plains and palm-fringed shores rocked and were unsteady. The Gods conferred. They decided the answer lay in placing a mountain upon the island to balance, calm and soothe it. And so this they did. Thereafter happiness reigned on the island and all was at peace. The mountain was called Great Mountain – Gunung Agung – and the island was Bali.

With its dense tropical jungle, lush green rice terraces, brooding mountains and lakes mirroring temples, Bali is claimed by many to be the last paradise on earth. Although arguably now over-visited, its accessibility makes it a perfect prelude to a cruise amongst some of the most enchanting islands in the world.

A DREAM SHIP

Few of us, sadly, in these straitened times, can maintain a fully crewed ocean-going yacht, but we can dream of running away to sea on a pleasure ship. As the humidity of the frenetic dock at Bali was replaced by the cool elegance of the oval-shaped reception hall of the *Silver Cloud*, my dream voyage quickly became reality.

A glass of chilled champagne, deftly served by impeccably attired Austrian stewards and a

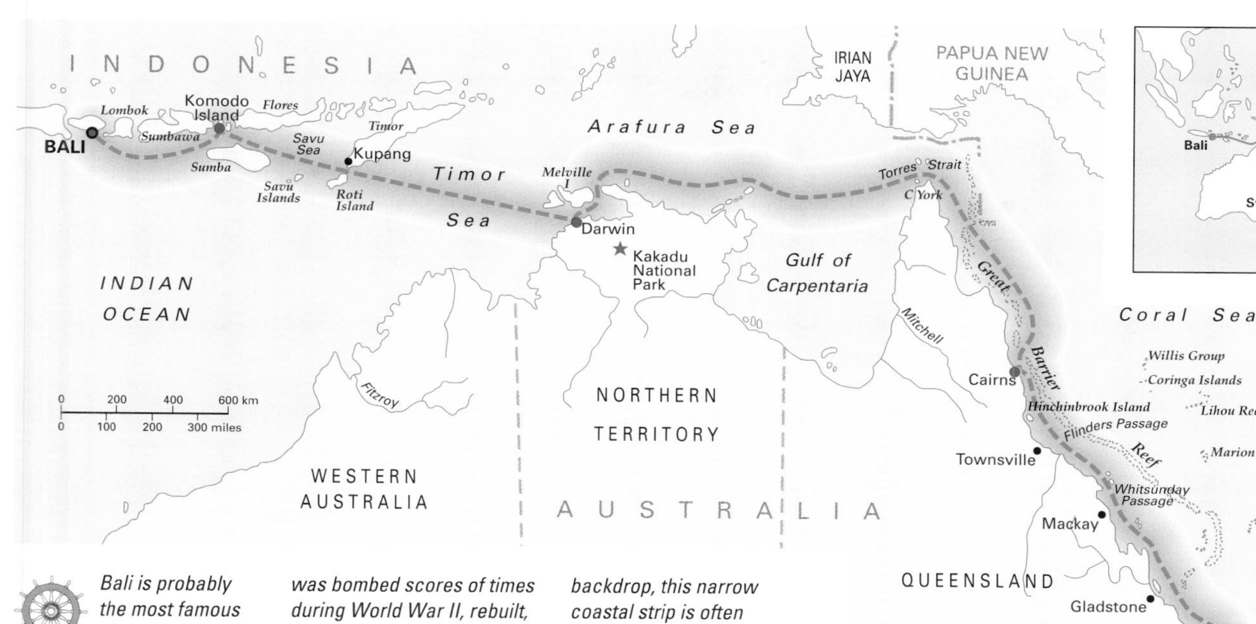

Bali is probably the most famous of the 13,600 islands of the Indonesian archipelago, the world's largest island group, and the Balinese have a vibrant culture whose beliefs, rites and festivals have long fascinated overseas visitors. Much of this culture is bound up with the island's religion, an ancient form of Hinduism, to which most of the 3 million inhabitants adhere.

Komodo Island is home to Varanus komodoensis, *or the Komodo Dragon – a protected species on the island that can weigh up to 300lb (135kg). They are sleek, fast runners with tapered heads, long necks, strong legs and long, powerful tails. They prey mostly on small mammals, birds and carrion.*

Darwin is a young, prosperous city that has survived more than its fair share of catastrophes. It was bombed scores of times during World War II, rebuilt, then wiped out by a cyclone in 1974. Today, modern Darwin is centred around Smith Street Mall, a pedestrian-only shopping centre in the heart of the restored city. For many visitors, this is their first encounter with Aborigines, who often congregate in the Mall.

World Heritage Listed rainforests fringe the tropical city of Cairns in North Queensland and getting an aerial view of this beautiful area couldn't be simpler as Skyrail, the world's longest gondola cableway, stretches 5 miles (8km) across the rainforest canopy. From this vantage point you can see, hear, smell and become part of this tropical environment.

Just south of Brisbane lies Australia's premier winter playground – the Gold Coast. With its scenic mountain backdrop, this narrow coastal strip is often compared with Miami due to its high-rise hotels and apartments ranged in phalanxes along the seashore. Altogether more restrained, the Sunshine Coast, lying to the north of Brisbane, offers some of the finest surfing in Australia.

Sydney is the most remote of the world's great cities but nonetheless it is a great tourist attraction with sights ranging from the Rocks area, where in 1778 a working party of convicts was landed to clear the ground and establish the first British settlement in Australia, to the Sydney Tower, the highest building in the southern hemisphere at 1,000ft (304m).

BELOW, *one of the grand suites, with the bonus of a different view every day*

BELOW, *the sleek lines of* Silver Cloud. *Few other cruise ships offer comparable space, luxury and attention to detail*

welcome as warm as the Balinese sun reassures even the most dubious traveller that the next 14 nights will be as magical as anything the 'island of the Gods' could conjure up.

As the sleek white ship cast her lines and released my last grip on this island paradise, the myriad of whispering coconut palms and brilliant tropical blossoms that form a panoply around Gunung Agung began to disappear over the aft-deck. This seemed like an appropriate time to explore my new peripatetic nirvana.

I discovered richly grained wood cabinets with pastel spreads and drapes; a private lounge leading on to a private veranda; a closet that was twice as large as my own back home and a generously proportioned bathroom filled with expensive toiletries and acres of fluffy towels. Attention to detail was omnipresent, from the personally printed stationery to the television screen positioned strategically for bedtime viewing. Within minutes of exploring the six passenger decks awash with gleaming glass and scrubbed teak it became evident that this was less of a cruise ship, more an exclusive private club.

This is a world where lobsters are flown in for dinner, of water-skiing and windsurfing, of elegant dining in the company of sophisticated friends and a level of pampering long gone from most five-star hotels.

My fellow travellers – just 280 in all – exuded a aura that spoke of sophistication, discerning taste and a wealth of experience in the wonderful world of ocean cruising.

THERE BE DRAGONS

Peaceful is an apt description of Komodo Island, but not of its most famous attraction. An early rise was rewarded as the lightening sky brought an ethereal monochrome grey into life over the bay and mountains of this spice island, more famous for its four-legged residents than its commodities. Komodo Island is a national park and home to the largest prehistoric monitor lizard in the world – the Komodo dragon. These amazing creatures that have been extinct everywhere else in the world since the Jurassic age grow up to 10ft (3m) long and have the run of the island. The excursion to observe these fierce predators, who can kill with one bite due to a strain of virulent bacteria in their mouth, is just about as other-worldly an encounter as it is possible to experience on a cruise today.

BELOW, *not quite as benign as he looks – a bite from a Komodo dragon can be lethal*

That evening, my second on board, was the first formal night of the cruise. Sequins and sapphires rubbed shoulders with tuxedos and topaz. The handsome Captain di Palma assured us that during our cruise we would enjoy calm seas and tantalising glimpses of picture-perfect ports far removed from the regular haunts of other, less refined, cruise destinations – like his ship, he didn't short-change on style.

The restaurant almost defies description; plush seats and gleaming table settings making a perfect stage for the nightly entourage of passengers in their tropical finery as they colourfully dress the room. That night I dined with a charming couple from Palm Beach, inveterate cruisers both. Barbara was taking her forty-third cruise and Allen his sixth. On *Silver Cloud* there are no stuffy – or uncomfortable – table assignments, you dine with whom you want.

A post-prandial promenade around the open decks in a balmy breeze under a sky lit by a galaxy of stars was to become a nightly routine of indescribable magic. Later, in the privacy of my own suite sanctuary, the ship's gentle motion proved to be a powerful hypnotic.

The following morning I was invited to visit the futuristic bridge. In front of the gently pitching bow there was just a huge arc of space defined by the straight line of the horizon, a rigid and unyielding circle unbroken for 360 degrees that separated the blue sky from the indigo sea.

After an al fresco breakfast in the terrace café followed by a lazy Jacuzzi on deck, a steward neatly arranged a thick towel for me on an oh-so-comfortable sun-lounger.

DREAMTIME

Daybreak came and we arrived at Darwin, the capital and only real city in Australia's Northern Territory – a region of the Outback, an area which covers about one-sixth of the entire continent. I made an early start to join my fellow passengers on an excursion to the Kakadu National Park. Some of the paintings we see on Kakadu's rock ledges have been there since Europe's Palaeolithic cave paintings were created. The Aborigines believe that all the earliest paintings were not in fact painted by man at all, but by Mimi spirits from the Dreamtime which still live inside the rocks and come every evening to hunt and paint. Some paintings hold the mysteries of the Dreamtime and the time of creation, as well as, the Aboriginal elders believe, the future of Kakadu.

This vast natural wonderland is a place of old magic and the most immediate, breathtaking beauty. Past every tree several of the 275 species of bird appear while beneath the waters swim many of the 50 species of freshwater fish. Throughout the park there are 51 species of mammal, 75 species of reptile, 25 kinds of frog and an estimated 10,000 species of insect. The abundance of fauna and flora in Kakadu is unrivalled anywhere in Australia and equalled only in a very few places around the world.

That night, as we sailed around the northern shores of Australia, I joined a delightful group from Sydney for the Broadway-style production show in the Venetian show lounge. The hits of Cole Porter and Irving Berlin brought a sense of the familiar to conclude a day that was suffused with the surreal.

As *Silver Cloud* followed an easterly course across the Gulf of Carpentaria the following day, clouds as dark as night and bursting with rain lay on the horizon. Within minutes, flickering tongues of lightening illuminated the decks and sent sun-seekers scurrying for sanctuary. As quickly as they had appeared the storm clouds receded behind the ship's white wake.

THE GREAT BARRIER REEF

Rounding Cape York and across the Torres Strait, halfway through our voyage, the centre of activity was the show lounge where an expert lecturer on the natural history and geography of the Great Barrier Reef gave us an insight into that great biological manifestation that laid ahead. Although the talk was interesting and informative, nothing could prepare us for the stunning reef itself – Australia's most wonderful spectacle and the world's largest living phenomenon.

Stretching for 1,240 miles (1,984km) off the north-eastern coast of Australia, the term 'Great Barrier Reef' is somewhat of a misnomer. The reef consists of 2,100 separate coral outcrops, scores of vegetated cays, and 540 lush islands. Whilst it is the habitat of 2,000 species of fish and thousands of different molluscs, sponges, worms and crustaceans, it's justification as the world's largest living thing is by merit of its composition of over 500 different varieties of coral.

ABOVE, *Aboriginal rock art at Nourlangie Rock in Kakadu National Park*

BELOW, *Cairns marina set against the backdrop of Trinity Bay*

PRACTICAL INFORMATION

■ *Silver Cloud* began cruising in 1994 and her sister ship, *Silver Wind*, entered service in 1995; together they comprise the present fleet of the ultra-deluxe Silversea Cruise Line, based in Fort Lauderdale. Each of the 16,800-ton ships carries a maximum of 296 guests in 148 suites, 75 per cent of which feature a private veranda.

■ The Bali to Sydney cruise is part of the line's annual four-month season in Oriental and Antipodean waters and the ports of call include Komodo Island in the Indonesian archipelago, Darwin in Australia's Northern Territory, Cairns in the north of the Australian state of Queensland, Brisbane, Queensland's capital city, and Sydney.

■ Throughout the year both *Silver Cloud* and *Silver Wind* also cruise South America and the Caribbean, the Mediterranean, the British Isles, the Norwegian Fjords and the Baltic, as well as Canada and Colonial America.

■ Silversea Cruises European office is: 77/79 Great Eastern Street, London EC2A 3HU. Tel: 0171 613 4777. In the USA call: 800 722 9055, ext 112. From January to May either *Silver Cloud* or *Silver Wind* sails on a series of 13- and 14-night cruises out of Auckland, Bali, Bangkok, Cairns, Hong Kong, Singapore and Sydney. Passengers should check visa requirements, especially for Vietnam and Australian cruises, and it is always advisable to travel with US dollars.

■ These cruises are timed for optimum weather conditions which means that for most destinations this is also the most

On 11 June, 1770, Captain Cook's barque HMS *Endeavour* was gored by a hidden coral outcrop off Hinchinbrook Island. Captain di Palma was taking no chances as he navigated *Silver Cloud* south through the Coral Sea prior to taking the inside passage across tranquil emerald green seas before making landfall at Cairns.

Set in the dazzling waters of Trinity Bay, Cairns is known as 'Queensland's Playground'. I joined the rest of *Silver Cloud's* complement of intrepid voyagers for a complimentary 'Silversea Experience' on board the *Ocean Spirit*. This, the world's largest sailing catamaran, sailed eastwards to fascinating Michaelmas Cay where an unforgettable day of snorkelling, scuba diving and windsurfing was enjoyed on this pearl-shaped island surrounded by pure white sandy beaches and water so blue and clear it seemed we were not in paradise but in a dream of paradise.

As *Silver Cloud* navigated the spectacular Whitsunday Passage and crossed the Tropic of Capricorn the following day, an Italian buffet groaned with attractive antipasti while the air was heady with the smell of freshly pressed garlic simmering in a variety of pasta dishes, deftly prepared by the Italian head waiters.

I joined Alex and Mabel on the wrap-around promenade deck, just in front of the observation lounge as we sailed up the Brisbane River. This was a homecoming for this Queensland couple and they gave me a personal narration of the principal sights of this fresh, clean and thoroughly modern city. Brisbane, home to the 1988 World Expo, has recently shrugged off its epithet of 'branch office town'. Now, restored Victorian sandstone buildings housing international designer fashion shops nestle cheek by jowl with modern structures encompassing Australian opal stores and Aboriginal art galleries.

Sailing south on the final leg of our Antipodean odyssey, the pool deck was given over to an Australian 'barbie'. This informal party continued well into the night as we danced under an inky-blue ceiling studded with an infinity of twinkling constellations.

Just as Bali is the perfect overture to such a voyage, Sydney is the ultimate finale. The billowing white sails of the Opera House, the schools of yachts at Rushcutters Bay and the majestic iron arch-span of the Harbour Bridge ushered *Silver Cloud* to her berth alongside Circular Quay in the heart of the historic Rocks area.

This heavenly gateway to Australia proved to be a perfect place to harbour dreams. Our voyage from the 'island of the Gods' through sublime seas was over but, thankfully, every cloud has a silver lining, and ours was to discover this great city.

INSET, *stunning blues of Michaelmas Cay, one of the low sandy islands (cays) forming part of the Great Barrier Reef*

RIGHT, *the spectacular Whitsunday Islands, so named by James Cook because he reached them on that day, in 1770*

popular time for tourists. The benefit of Silversea's ships are that with a shallow draught of just 17ft (5m) they can cruise waters and rivers that other, larger ships cannot navigate – allowing them to tie up alongside a convenient downtown berth. With most arrival times scheduled for around 8am, cruise passengers usually beat the rush of hotel-based visitors to the principal attractions.

■ The programme of Silversea Cruises pre- and post-cruise land-stay options is a comprehensive one and, in keeping with the demands of their passengers, most of these include stays at some of the world's finest hotels.

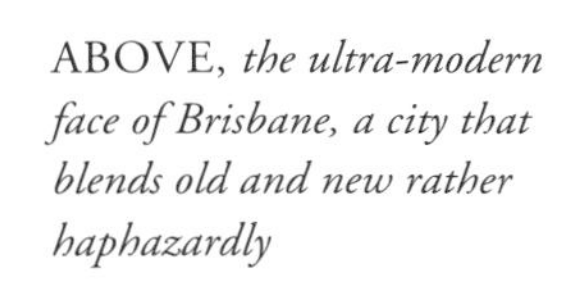

ABOVE, *the ultra-modern face of Brisbane, a city that blends old and new rather haphazardly*

BELOW, *Sydney Harbour, with Australia's most famous icon, the white-sailed Opera House*

Breaking Through to the North Pole

BOB HEADLAND

Even during the height of summer the Arctic Ocean is never less than half covered by many feet of pack-ice, penetrable by only the most powerful of ice-breakers. When not carrying passengers, the Russian-owned atomic vessel *Yamal* helps keep the North-east Passage open for shipping. For most of the journey to the North Pole the ship ploughs relentlessly through the ice, well north of any human habitation and with few signs of wildlife.

ABOVE, *a little Nenet girl wrapped up warmly against the bitter cold*

BELOW, *dwarfed by the awesome bows of a Russian ice-breaker*

Most of the voyages to the North Pole begin from Kirkenes in the far north of Norway, where there is daylight 24 hours a day during the summer months.

We made an early start for Murmansk, taking a hydrofoil (which felt like riding an elephant, so seasickness precautions may be useful) into the centre of the port, where lengthy frontier formalities were completed. We were then taken on a bus tour of the city before boarding the ice-breaker *Yamal,* to which we were welcomed in traditional Russian fashion with bread and salt.

Several sister ice-breakers were also docked at Murmansk, including *Arktica*, the first surface ship to reach the North Pole, in 1977. This was just under 70 years after the first overland expedition made it. After casting off it took a couple of hours to reach the end of the Murmansk fjord, a journey which provided amazing views of the Russian fleet.

On the first day we were given preliminary briefings and lifeboat drill, and a series of lectures began which contributed greatly to the understanding and enjoyment of the voyage.

Yamal is huge and on exploration discoveries such as a swimming pool, sauna, library, bar, gymnasium and theatre were made. Accommodation, although quite adequate, is not luxurious. The day ended with a special dinner attended by the captain and many of his officers, which provided an excellent opportunity to meet those operating the ship and find out how she worked.

As we journeyed on we glimpsed an Arctic fox well to the west of Franz Josef Land (Zemlya Frantsa-Iosifa) and later three accommodating polar bears almost posed for photographs. The ship stopped and their curiosity overcame any fear

Some voyages reach the North Pole and return to Murmansk while some go all the way to the Bering Strait before returning. Several of the truly fascinating and almost unknown series of Arctic archipelagos are visited. These are replete with wildlife, historical sites and have much of scientific interest. The 191 separate islands of Franz Josef Land (Zemlya Frantsa-Iosifa) have some of the most spectacular bird colonies anywhere, especially at the huge volcanic plug called Rubini Rock. Mammoth remains are common on the New Siberian Islands, while the musk ox presently may be seen on Wrangel Island. Severnaya Zemlya, the last significant area of the Earth to be discovered, abounds in ice domes and geological Arctic features. Large populations of polar bears, seals and walrus inhabit the entire region. The ice-breaker is able to approach them quite closely.
The islands have served as bases for many attempts on the North Pole. Some areas, especially Cape Flora (Mys Flora), have a surprising wealth of wild flowers, spectacular mosses and lichens, and other vegetation, at the appropriate season. Occasionally stops are made on mainland Russia, and villages, sometimes of predominantly indigenous populations (Nenets, Chuchi, and others), may be visited. Sometimes a reindeer herders' encampment may also be seen. Even remains of the gulag system may be visited.
Further north the only habitations are remote polar stations with a few meteorologists in residence, and even these are dwindling in number as many are being closed through economic necessity.

LEFT, *this exile cabin on the New Siberian Islands is now a trappers' refuge*

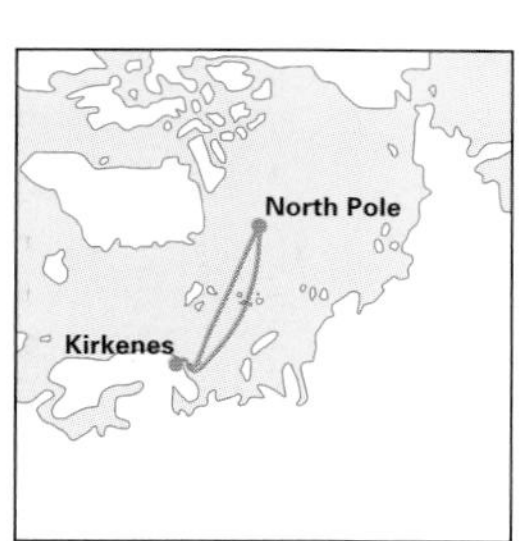

they may have had – they approached within 165ft (50m), possibly encouraged by the thought of all that food lining the forward rails?

Apart from watching our progress through the increasingly thick ice, there was much to do aboard. An interpreter and language teacher from Murmansk offered lessons in Russian, which proved useful in deciphering the alphabet, and tours of the control room were also arranged – there were no restrictions on photography but dire threats were issued to anyone tempted to start pushing buttons or pulling levers. Leaded-glass ports allowed us to watch the two atomic reactors producing steam to power the engines.

A Bird's-eye View

As the ice thickened two helicopters were used to survey the best way ahead and flights were arranged so that passengers could see the ship working its way forward. A small conventional helicopter provided the better views, but the larger one with two opposed rotors on concentric shafts and no tail rotor was more comfortable.

Despite the difficulty in passing through the ice, it was safe to 'land' on and after lunch the following day the gangway was lowered and most passengers went 'ashore'. Summer, indicated by the amount of slush and surface water around, was advancing and it was interesting to recall that about 8,224ft (2,500m) of water lay just below the ice on which we stood. The ever present possibility of attack by polar bears was countered by two

ABOVE, *the dazzling white fur of the Arctic fox stands out even against the expanse of snow*

riflemen stationed at the outer limit of the walking area, which had been previously inspected for safety by prodding cracks and ponds with a large ice-chisel. While we were stationary, surprisingly large numbers of sea-birds, including the uncommon ivory gull, flew around the ship on the look-out for food.

Towards evening the weather improved, allowing passengers to land from the helicopter on the ice ahead and to one side of the ship: the sight of the ship passing by, forcing vast blocks of ice aside, was spectacular. Another Arctic fox was sighted, this one's pure white winter coat in the process of turning to its grey summer livery, thus making it slightly easier to see.

Later in the voyage progress became increasingly difficult. We encountered ice up to 10ft (3m) thick, and on many occasions the ship backed and charged again. Getting through was never a problem, but our speed was considerably reduced both by the back-and-run operations and by the rather sinuous course taken in order to find the path of least resistance.

After a while the going became rather easier and steadier. Meanwhile, preparations were being made for the arrival of HM King Neptune, Tsar of the Five Oceans, which provided an unsurpassed opportunity for a party. Held on the upper aft deck with the temperature just above freezing, it was attended by the king's retinue, the captain, many officers, a large proportion of the crew and all passengers. There was an excellent barbecue and although language was a problem, large quantities of beer (*pivo*), wine (*vino*) and vodka greatly helped communication.

We continued north. At these latitudes navigation becomes complicated (longitudes converge amazingly). So, to be certain of reaching the North Pole, a reconciliation and calibration of the satellite and inertial navigation systems was needed.

ABOVE, *looking out across the icy wastes towards Wrangel Island*

Progress became more difficult and much overturned and broken sea-ice was seen around the ship, some more than 10ft (3m) thick. Progress continued fitfully during the night, as did any practicability of sleeping as sometimes the air-bubble system was used to help break the ice and in some parts of the ship this sounded like the engines of a large jet aircraft about to take off.

On Top of the World

As we ploughed on, anticipation of reaching our goal rose. Few birds or mammals survive this far north and the only noticeable wildlife was the occasional patch of marine algae in the lower strata of the pack-ice. Next afternoon the weather was not particularly good; there was little wind and a fog reduced the visibility to less than 658ft (200m). Sky, cloud, mist and sea-ice all merged into one dense, white, all-enveloping blanket.

Yamal made steady but rough progress through the ice during the night. At about 6.30am the announcement that we were fewer than 14 nautical miles (25km) from our destination was made. Virtually everyone gathered on the bridge and an exception to the rule of no beverages was made as champagne was served. Shortly after 8am the ship was within several hundred yards of the 90°N position. Much careful navigating, measurement, checking of instruments and manoeuvring of the ship ensued before the captain was able to confirm that at 9.20am Moscow time (5.20am GMT) the ship was on the North Pole. Appropriate festivities began and continued for the rest of the day as light snow began to fall. Although it became heavier, by evening it had entirely abated and a clear, magnificent Arctic evening – the sun well visible overhead – emerged. Perfect.

The next task was to find a large, strong floe on which to hold a celebratory party. Those in the

ABOVE, *suitably protected from the weather, celebrations begin at the North Pole*

sound came when a propeller blade sliced through a particularly large block of ice.

One afternoon we visited a Russian craft shop where we found not only the expected postcards etc, but also a large number of lacquerware, stone, leather and woven items.

As *Yamal* approached Franz Josef Land the lectures concentrated on what we were to encounter there – the birds and mammals, geological and glaciological landforms. An early evening session of 'any questions' was a useful chance to debate some of the issues raised.

During the night the pack diminished in thickness and cover, with only an occasional thump to disturb our sleep, and by 6am the ship had reached the northern entrance to Cambridge Strait. Over half the strait was covered with pack-ice, and footprints of polar bears, plus several other animals, could be seen across the snow layer; occasional seal holes were visible too. The low-lying land and raised beaches on one side contrasted with the ice cap and glaciers entering the sea on the other. Both were beautifully illuminated by the clear sun but a low mist closed in as we reached the western

immediate vicinity were too small and covered with a lot of deep soft snow – as well as being churned up by the ship during her positioning exercise. After moving slightly south towards Canada a suitable place was found and by 12.30pm most passengers were on the ice. The sense of occasion was reinforced by speeches, 'North Polar Grog', a barbecue, music, a lot of pyrotechnics, as well as various other activities, including the waving of flags of the various countries represented aboard – Austria, Belgium, Canada, France, Germany, Portugal, Russia, Sweden, the United Kingdom and the United States.

South, the Only Option

After several hours on the ice we all reboarded the ship. The barbecue and even the Pole itself were hoisted aboard and we set off for Franz Josef Land. Owing to the previous manoeuvrings to find a safe floe this course took us over 90°N again, with much hooting from the ship's fog horn (two visits for the price of one). The great improvement in visibility and partly clear sky gave some of the best opportunities for helicopter sightseeing towards the end of an amazing day.

Yamal continued south (of course). During the return voyage we again encountered the particularly heavy ice which had caused some delay on the way north. Once again the sound of the compressors for the air-bubbling system pervaded the ship. On examining the zone where the ice meets the hull the effects of the air were often spectacular. A thick section of ice would occasionally burst out with a foamy geyser as it was lifted and crushed. The big bangs from large floes forced beneath the hull were quite common while we penetrated this patch of the ice. Occasionally a resounding ringing

BELOW, *a welcome burst of colour from flowering Arctic poppies*

end; nevertheless we were optimistic that it would burn off later in the day.

The fog did indeed clear, with first the tops then much of the lower regions of Bell Island (Ostrov Greem-Bell) and Mabel Island coming into view; by late afternoon one of the most magnificent days yet had developed. Helicopters were soon in the air and after a flight over some of the most spectacular scenery in the archipelago we landed near the hut set up for Benjamin Leigh Smith's expedition aboard *Eira* from Britain in 1881. Considering the hut's age, it was in amazingly good condition.

The brilliant weather held, and landings continued after dinner, by which time the ship had reached a station off Cape Flora (Mys Flora) on Northbrook Island (Ostrov Nortbruk). The cape was aptly named: Arctic buttercups and purple saxifrages were in flower, and the large flower buds of the Arctic poppy seemed ready to burst open any day. Mosses and lichens were particularly abundant and we had to tread carefully to avoid damaging them. Enormous numbers of sea birds nested in the cliffs and just before our departure more than 20 white beluga whales (including several dark-grey calves) were seen nearly a mile off-shore.

The weather held and *Yamal* continued along the British Channel (Britanskiy Kanal) to reach Cape Norway (Mys Norvegiya). Fortunately the sea was calm and the shore almost unencumbered with ice so we were able to land by Zodiac. We saw the stone hut where Fritjof Nansen and Hjalmar Johansen wintered under arduous conditions, including the drift-wood roof beam and bones of bears, walruses and seals they killed to survive.

Rubini Rock, on Hooker Island (Ostrov Gukera), is of special interest due to its amazingly convoluted columnar basalt, flourishing vegetation (moss and lichens) as a result of fertilisation by guano, and spectacular bird life – the site houses tens of thousands of both thick-billed murres and black-legged kittiwakes: the smell is memorable.

By now the voyage was drawing to a close and our last evening aboard the ice-breaker was marked by a special Russian dinner. Next morning we were back among trees, buildings and ships; we had reached Murmansk harbour.

LEFT, *after an eventful journey, returning to the relative normality of Murmansk harbour*

Practical Information

- Several companies advertise voyages on the Arctic Ocean but only those chartering atomic ice-breakers can take you to the North Pole. Apart from trekking, or going by submarine, the only alternatives are aircraft, which can only make brief visits and must keep their engines running. The principal company involved is Quark Expeditions, 980 Post Road, Darien, Connecticut, USA 06820. Tel: 203 656 0499, Fax: 203 655 6623.

- For logistical reasons, such voyages are made only during summer, late July to early September, when there is perpetual daylight. As many as three trips usually take place during this period. The vessel most commonly used is *Yamal*, the world's most powerful ice-breaker (75,000hp), which has broken pack-ice up to 20ft (6m) thick.

- Cabins have private facilities and are mainly two-berth; some staterooms are available.

- The headquarters is in Murmansk, so passengers will require Russian visas, and should leave sufficient time to obtain them.

Round the World on the QE2

GARY BUCHANAN

On any circumnavigation, *Queen Elizabeth 2* , commonly known as *QE2*, visits a variety of sun-drenched ports, including New York, Honolulu, Singapore, Bombay and Cape Town. For passengers who embark at Southampton shortly before Christmas, the full world cruise lasts for around 120 days, but it is not until Fort Lauderdale that the real global odyssey gets under way in earnest. And this the section of the trip on which we shall concentrate, as the full journey would, obviously, command much more space than is available here and, indeed, could fill a whole book on its own.

ABOVE, *the Statue of Liberty, that quintessential symbol of America, holds out the promise of a great adventure ahead*

BELOW, *Rio stages the most spectacular carnival in the world – over four days of exuberant celebration*

Cruising has come a long way since 40 BC, when Cleopatra sailed up the Cydnus River on a barge laden with gifts to steal the heart of Mark Antony, sealing her fate as one of history's most notorious seductresses.

More than 2,000 years later, there can be few travel experiences more desirable, more monumental or more extravagant than a pan-global odyssey. World cruising as we know it today dates back to 1922, when the Cunard ship, RMS *Laconia,* made her first circumnavigation of the globe. To celebrate the occasion of the first Cunard Line world cruise, a silver cup was presented to the ship. Today this same cup graces the Samuel Cunard Key Club on board the *QE2*.

CARNIVAL CITY

Ever since its heyday in the 1950s Rio de Janeiro has conjured up images of hedonism and excess. Giant carnival floats full of gyrating, semi-naked bodies, surrounded by a sea of sequins and feathers, accompanied by samba bands parading through the streets; Eva Peron arriving with 100 suitcases at the Copacabana Palace Hotel where Orson Welles chose to throw the entire contents of his room into the pool – yes, Rio, the *cidade maravilhosa* (marvellous metropolis), has always had a reputation for bringing out the party animal in even the most staid of souls.

Quickly shaking off my jet-lag I throw open the shutters of my second-floor room at the recently revamped Copacabana Palace hotel to marvel at the expansive Copacabana beach, an olive stone's throw across Avenida Atlantica. I am instantly

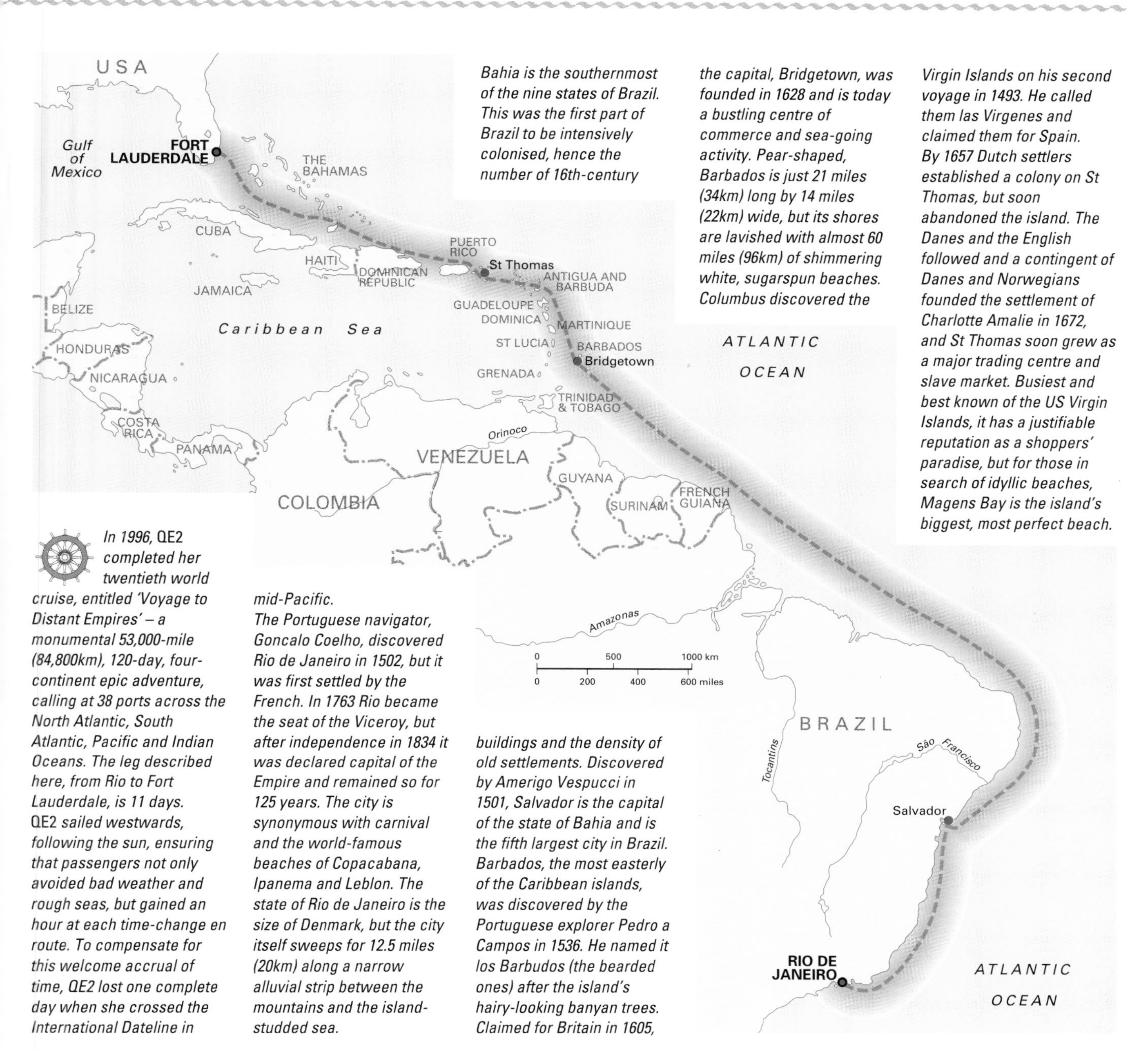

In 1996, QE2 completed her twentieth world cruise, entitled 'Voyage to Distant Empires' – a monumental 53,000-mile (84,800km), 120-day, four-continent epic adventure, calling at 38 ports across the North Atlantic, South Atlantic, Pacific and Indian Oceans. The leg described here, from Rio to Fort Lauderdale, is 11 days. QE2 sailed westwards, following the sun, ensuring that passengers not only avoided bad weather and rough seas, but gained an hour at each time-change en route. To compensate for this welcome accrual of time, QE2 lost one complete day when she crossed the International Dateline in mid-Pacific.
The Portuguese navigator, Goncalo Coelho, discovered Rio de Janeiro in 1502, but it was first settled by the French. In 1763 Rio became the seat of the Viceroy, but after independence in 1834 it was declared capital of the Empire and remained so for 125 years. The city is synonymous with carnival and the world-famous beaches of Copacabana, Ipanema and Leblon. The state of Rio de Janeiro is the size of Denmark, but the city itself sweeps for 12.5 miles (20km) along a narrow alluvial strip between the mountains and the island-studded sea.
Bahia is the southernmost of the nine states of Brazil. This was the first part of Brazil to be intensively colonised, hence the number of 16th-century buildings and the density of old settlements. Discovered by Amerigo Vespucci in 1501, Salvador is the capital of the state of Bahia and is the fifth largest city in Brazil.
Barbados, the most easterly of the Caribbean islands, was discovered by the Portuguese explorer Pedro a Campos in 1536. He named it los Barbudos (the bearded ones) after the island's hairy-looking banyan trees. Claimed for Britain in 1605, the capital, Bridgetown, was founded in 1628 and is today a bustling centre of commerce and sea-going activity. Pear-shaped, Barbados is just 21 miles (34km) long by 14 miles (22km) wide, but its shores are lavished with almost 60 miles (96km) of shimmering white, sugarspun beaches.
Columbus discovered the Virgin Islands on his second voyage in 1493. He called them las Virgenes and claimed them for Spain. By 1657 Dutch settlers established a colony on St Thomas, but soon abandoned the island. The Danes and the English followed and a contingent of Danes and Norwegians founded the settlement of Charlotte Amalie in 1672, and St Thomas soon grew as a major trading centre and slave market. Busiest and best known of the US Virgin Islands, it has a justifiable reputation as a shoppers' paradise, but for those in search of idyllic beaches, Magens Bay is the island's biggest, most perfect beach.

LEFT, *built in 1969 and refitted many times since, the* QE2 *is one of the most splendid cruise ships afloat today*

ABOVE, *the magnificent view of Rio and Sugar Loaf Mountain that can be had from Corcovado*

BELOW, *prestigious hotels and a busy expressway line Rio's bustling Copacabana beach*

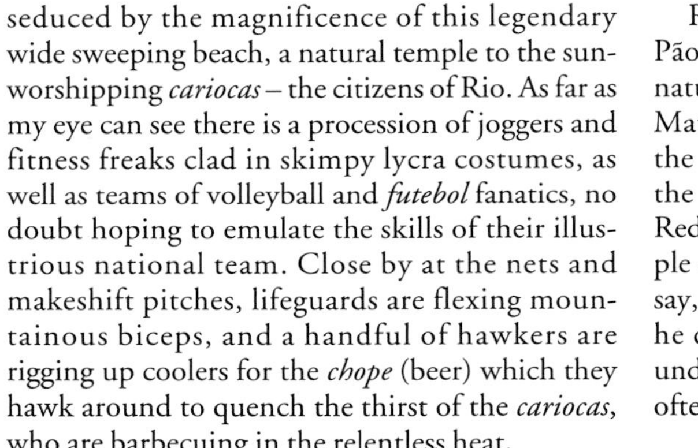

seduced by the magnificence of this legendary wide sweeping beach, a natural temple to the sun-worshipping *cariocas* – the citizens of Rio. As far as my eye can see there is a procession of joggers and fitness freaks clad in skimpy lycra costumes, as well as teams of volleyball and *futebol* fanatics, no doubt hoping to emulate the skills of their illustrious national team. Close by at the nets and makeshift pitches, lifeguards are flexing mountainous biceps, and a handful of hawkers are rigging up coolers for the *chope* (beer) which they hawk around to quench the thirst of the *cariocas*, who are barbecuing in the relentless heat.

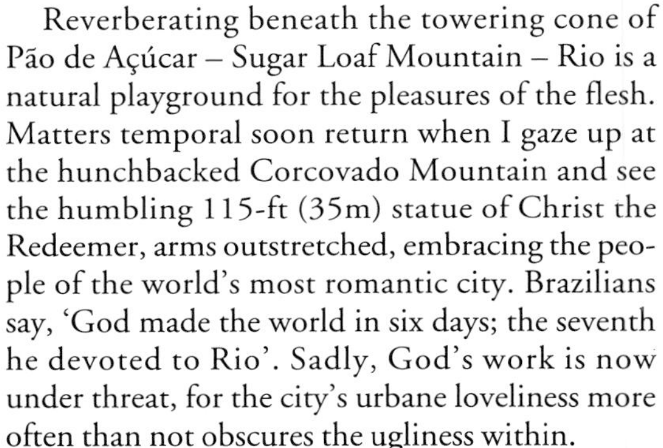

Reverberating beneath the towering cone of Pão de Açúcar – Sugar Loaf Mountain – Rio is a natural playground for the pleasures of the flesh. Matters temporal soon return when I gaze up at the hunchbacked Corcovado Mountain and see the humbling 115-ft (35m) statue of Christ the Redeemer, arms outstretched, embracing the people of the world's most romantic city. Brazilians say, 'God made the world in six days; the seventh he devoted to Rio'. Sadly, God's work is now under threat, for the city's urbane loveliness more often than not obscures the ugliness within.

Ultimate Luxury

The late afternoon sun exaggerates the intensity of the 'Cunard Red' funnel. After some brief formalities at the cruise ship terminal carried out by several smart officers from the purser's office, I find myself standing underneath the vast hull of this 70,327-ton leviathan. As *QE2's* superstructure soars high above me, the immediate thoughts passing through my mind are how big, blue and beautiful this ship is. I embark via a gangway leading to a massive steel door set into the side of the ship; above it hangs a sign inscribed with the two words that for me conjure up the very essence of world cruising – 'Welcome Home'.

Settling into my cabin on One Deck, I am immediately struck by the aura of decadence that envelops me in this delightful stateroom, fitted out in richly ingrained woods concealing more closet and drawer space than would appear seemly. *QE2* emerged from the swinging 60s, survived the economic shockwave of the 70s and sailed through the enlightened 80s, to come of age in 1988. With diesel motive power replacing steam, she emerged full of youth and vitality to sail with renewed vigour into the exciting 90s. With creative designs incorporated during her many refits, this gracious dowager duchess appears to be in the prime of her life as she sails towards the next millennium – vast, confident and wonderfully stately.

During cocktail hour, *QE2's* nine generators are cajoled into life and, with the assistance of two powerful tugs, we inch away from our Brazilian berth. Making a pronounced turn to starboard, this Titan of the seas – ablaze with 1,000 lights – sails under the flight path of commuter aircraft on their final approach into Santos Dumont Airport. A few minutes later Sugar Loaf Mountain, which marks the entrance to Guanabara Bay, appears on the starboard bow. I watch, transfixed, as a cable-car descends from the summit to the halfway point at Morro da Urca. Suspended on its filigree wires it is silhouetted against the night sky, as dark as a preacher's gown.

I postpone any thoughts of an early dinner as the vast expanse of Copacabana Beach hoves into view. Lined with international hotels, the Avenida Atlantica shimmers like a halo around the floodlit beach, while in the distance a backdrop of a myriad lights illuminates the night sky. The whole scene is crowned by the illuminated statue of Christ the Redeemer bestowing a pious transcendental blessing across this pulsating city. Rio de Janeiro is truly one of the world's most spectacular cities.

As the ship turns to port and heads out to sea, my thoughts turn to dinner. The exclusive Queen's Grill is the preserve of just 240 passengers who occupy the most prestigious cabins and penthouses on board. The gold-leaf ceiling of the lower, central section reflects the specially ordered flambée dishes, while the smart blue and cream fabrics are set off against the pale walnut of the surrounds. Crowning the room is the royal coat of arms, resplendent in polished oak.

After dinner I head for the Crystal bar. This large room, stretching the entire width of the ship, has an art deco-style semi-circular wooden bar adorned by a glass bas-relief of 'Winged Horse and Clouds', similar to that which graced the *Queen Elizabeth*. The harpist playing gentle melodies adds to the all-consuming air of decadence that pervades every inch of *QE2's* ten passenger decks.

Ahead of me lie two sea days – time enough to get acquainted with this vast city at sea. I took my deck-plan in my hand and set off to cover the whole ship. *QE2* is designed on the plan of a three-dimensional grid; decks horizontal – as one has the right to expect, the stairs vertical. It's a simple system but some passengers continually seemed lost. I come across several people, who by all accounts had joined the ship several weeks before me, holding hands with the person in front like Spahis in a sandstorm, walking up and down, round and round, eating at the first restaurant that would welcome them in – the Handsels and Gretals of the oceans. 'It doesn't matter', they assure me, 'we love it, it's the only way to travel.'

There are five restaurants on board, seven if you count the buffet-style Lido and hamburger haven of the Pavilion. The Queen's, Princess and Britannia grill rooms are small and intimate while the lavishly appointed Caronia and Mauretania restaurants rekindle images of the grand saloon. There is also a plethora of bars ranging from the Golden Lion pub, serving draught beer, to the

ABOVE, *by night and day, the statue of Christ the Redeemer watches over the city of Rio de Janeiro*

BELOW, *all quiet at the Golden Lion pub, one of several bars on board the* QE2

ABOVE, *the harbour at Salvador, capital of the state of Bahia*

elegant chart room, complete with a glass panel mounted behind the bar that depicts the north Atlantic and the Great Circle routes navigated by *QE2* during her many transatlantic crossings.

That evening, passengers conform to the ancient rite of sea passage and dress for dinner. I take my seat at the convivial bar in the Queen's Grill lounge and within seconds become engrossed in an amusing conversation with an affable man of indeterminate years from Houston. Holt has been on board since Singapore and is a marvellous source of gossip and scandal.

Ashore at Salvador

The sight of *QE2's* colossal anchor crashing into the sea off Salvador brings home the vastness of this great vessel. Weighing 12.5 tons each and attached to 1,808-ft (550m) cables whose massive links are each 4in (10cm) in diameter, the two mighty anchors – one forward, one aft – hold the ship fast, while tenders ferry passengers ashore.

Salvador, the capital of Bahia, is one of Brazil's most historic cities, with a wealth of colonial architecture; it is also dubbed 'Africa in exile'. The centre of the city is divided into two levels, *Cidade Alta*, or Upper City, where the historical centre lies and *Cidade Baixa*, or Lower City, where passengers are deposited on the landing stage and haggle with local taxi drivers to take them on sightseeing tours.

Dominating the Praca Municipal is the old Paco Municipal (Council Chamber) dating from 1660. I walk north from this square in search of a church named Terreiro de Jesus, leaving the panoramic view of the bay behind. Across Praca da Se, hanging with mimosa and set alight with the vivid foliage of the flamboyant trees, I can see my goal. Built in 1692 of Portuguese Lioz marble, the church of the Jesuits became the property of the Holy See in 1759 when the Jesuits were expelled from all Portuguese territories.

The building's façade is one of the earliest examples of baroque in Brazil, an architectural style which was to dominate the churches built in the 17th and 18th centuries. The interior is particularly impressive, its vast vaulted ceiling and 12-sided altars in baroque and rococo style framing the main altar which is magnificently leafed in gold.

The cooling air-conditioning on board *QE2* is a welcome relief from the oppressive humidity of tropical Brazil and I decide to visit the library and discuss my morning's excursion with the world's first librarian at sea, June Applebee. She points me towards some of the 6,000 volumes and I sink into a soft, squashy chair to await the time-honoured tradition of afternoon tea.

CRUISING THE CARIBBEAN

After four magical days at sea, where time is marked daily at noon with a long, melancholy blast of the ship's whistle, *QE2* crosses the equator to arrive at the Caribbean island of Barbados. Here a steel-band welcomes the 1,400 passengers who disgorge down the gangway to head off for island tours or hail taxis for the short ride into Bridgetown, intent on some serious duty-free shopping. I opt for a more leisurely day relaxing by the pool at the pastel-pink palace of the Royal Pavilion hotel on the 'platinum coast'.

The penultimate port before Fort Lauderdale is the tax-free Mecca of St Thomas in the US Virgin Islands. Lying at anchor off Charlotte Amalie – the bustling capital of St Thomas, *QE2* looks majestic: with her dark royal blue and snow-white hull, bisected by a gold-edged racing stripe, she is the epitome of classical ship aestheticism. I pass by

ABOVE, QE2's *well-stocked library would put many to shame*

BELOW, *Bridgetown's inner harbour is known as the Careenage*

PRACTICAL INFORMATION

■ Cunard reservations offices can be contacted at: South Western House, Canute Road, Southampton. SO14 3NR. Tel: 01703 716605, 555 Fifth Avenue, New York, NY 10017 2453. Tel: 800 7 CUNARD and Level 1, 146 Arthur Street, North Sydney NSW. Tel: 2 9956 7777.

■ World cruises always take place during the northern hemisphere's winter months of January through April and head for the sun-blessed destinations of the tropics, the Antipodes, Asia, South Africa and occasionally South America.

■ World cruises on board *QE2* generally follow a familiar routing, calling at principal cities on the four continents. Annual deviations to itineraries occur around Africa, when the 1,500-passenger liner either sails through the Suez Canal and into the Mediterranean or navigates around the Cape of Good Hope. At Cape Town there is another option of either heading on a northerly course, recreating a line-voyage to Southampton or heading west across the South Atlantic to Brazil. It was the latter option that QE2 undertook during her world cruise in 1996.

■ Every year during the world cruise, *QE2* transits the Panama Canal at least once. The Suez Canal, whilst exciting, has little of the drama of the Panama Canal, due largely to the absence of locks on the Egyptian waterway. On any circumnavigation, *QE2* normally crosses the International Dateline once and the equator four times.

RIGHT, *there are few more relaxing places than the white expanse of St Magens beach, St Thomas*

a plush jeweller's while, in the novelty shop, Holt is buying a fluffy pink pig to amuse our companions in the Queen's Grill.

Commodore John Burton-Hall's farewell gala cocktail party evokes a picture of an altogether more glamorous age, when the journey, not the destination was the important thing. Above all, during my ten-day sojourn I have come to understand that passengers have a preoccupation with *QE2*, partly because she is a conduit to memories of a lost era – a brief glimpse back into the remarkable world in which their forefathers crossed oceans in the grand days of travel, and partly because – as the saying goes – 'the best things happen at sea'.

No other modern cruise ship can compare with *Queen Elizabeth 2*; indeed, no passenger ship in the history of cruising can boast such a litany of superlatives – the flagship of Britain's Merchant Marine. 'Ship-shape and in Cunard fashion' is an appropriate epithet for this regal ship – the last in that great lineage of ocean-going liners. Today, as she matures like a fine wine, *QE2* is setting an historic course into the next millennium.

Passengers receive certificates to mark each event.

■ On-board accommodation ranges from split-level and duplex penthouse suites with hot and cold running butler service to somewhat cramped, but nonetheless cosy cabins in the lower echelons on Four and Five Decks. In all, there are 21 grades of cabins available. Despite the range of grades, QE2 is sufficiently egalitarian to operate as a one-class ship.

■ Grill room or restaurant allocation depends on cabin categories. Passengers occupying luxury penthouses and ultra-deluxe cabins on One and Two Decks dine in the Queen's Grill; those occupying deluxe cabins on One, Two and Three Decks dine in either the Princess Grill or the Britannia Grill; the Caronia restaurant – traditionally the main first class restaurant, is where passengers occupying cabins on Three and Four Deck dine, while those travelling in Four and Five Deck cabins dine in the Mauretania restaurant. All restaurants and grills operate on a one-sitting basis.

ABOVE, *Independence Arch stands proud in Bridgetown, Barbados*

LEFT, *a steel band provides traditional entertainment at the Plantation Show*

About the Authors

Gary Buchanan is a freelance travel writer based in Fife, Scotland. His interest in cruise travel spans 15 years and he has visited most of the world's most popular ports and many other interesting destinations, some of which are only accessible by ship. From floating leviathans to intimate yacht-like craft, he has crossed vast oceans and navigated remote waterways.

Born in London, Liz Cruwys received a PhD from the University of Cambridge, where she conducts research in marine biology. Author of four successful medieval mystery novels, she also contributed to the AA's *Houses of Treasure*, *Great Highways of the World*, *Natural Wonders of the World* and the *Road Book of Britain*, and co-wrote *Cathedrals of the World* and *Explore Britain's Castles* with her husband Beau Riffenburgh.

Ben Davies is a travel writer, journalist and photographer. Based in Bangkok, where he is associate editor for Asia Money and local correspondent for World Link, he has also contributed to newspapers and magazines ranging from the International Herald Tribune to the Traveller Magazine. He recently wrote, photographed and published *Isaan: Forgotten Provinces of Thailand*, and also contributed to *Great Highways of the World* and *Train Journeys of the World*.

Since spending her formative years on a Pacific beach in Australia, Fiona Dunlop has been fascinated by the tropical climes and cultures of the planet, despite long periods spent living in London, Paris and Italy. Besides covering contemporary arts and design for numerous magazines, she has also written several AA Explorer Guides, notably *Paris*, *Singapore and Malaysia*, *Indonesia*, *Mexico*, *Costa Rica*, *Vietnam* and *India*.

Peter and Helen Fairley are freelance travel writers with experience of more than 50 countries. Peter began writing in the 1950s, becoming an expert on space travel. In addition to writing 13 books, he has also made many broadcasts. His wife Helen is a freelance photographer who was involved for many years with the English Tourist Board.

Robert Headland is archivist of the Scott Polar Research Institute at the University of Cambridge. This gives him excellent resources for studying the Arctic and Antarctic – areas of which he has great practical knowledge. Since the breakup of the Soviet Union he was one of the first to reach previously prohibited parts of the Arctic, aboard Russian nuclear ice-breakers. He also lectures on historical geography.

Christopher Knowles has lived in France and Italy and travelled all over the world, from Albania to Zimbabwe, as a tour guide and writer. He spent a great deal of time in China and the former Soviet Union in the 1980s and has written guide books on Shanghai, Moscow and St Petersburg and China, as well as Florence and Tuscany and the Cotswolds.

Shirley Linde is a writer on science and travel for major magazines, and the best-selling author of 35 books on a variety of subjects, including *The Insiders' Guide to the World's Most Exciting Cruises*. She has received several service rewards in communications and is listed in *Who's Who in America* and *Foremost Women of the 20th Century*. She lives in the USA on the water near St Petersburg, Florida.

Beau Riffenburgh lived in Los Angeles before moving to Cambridge, where he received a PhD and where he edits *Polar Record*, the journal of the Scott Polar Research Institute. He has written numerous journal, magazine and newspaper articles, and 12 books, ranging from the history of exploration to American football. He was also co-author of the AA's *Great Highways of the World* and *Natural Wonders of the World*.

Anthony Sattin has been travelling in the eastern Mediterranean and Middle East since he was old enough to apply for a passport, and divides his time between London and Cairo. He has written several books, including a history of travellers in Egypt, and also discovered Florence Nightingale's lost letters from Egypt.

Ann F Stonehouse is an editor, writer, traveller and fiddler. She has contributed to the *Cambridge Guide to Literature in English*, and articles on Scotland to several books. She has a particular interest in John Buchan, and wrote the introduction and notes for the recent World's Classics edition of his novel *Huntingtower*.

Rob Stuart divides his working life between television and travel, and is currently working for Grundy Europe in Sweden. His first travel article, a piece on rowing the River Severn, was published in *The Daily Telegraph*. He is a regular contributor to the *Telegraph's* travel pages, and also writes for *The Guardian*.

A Royal Air Force wife, Mary Tisdall acquired her love of travel living in Jordan, Malta and Singapore. Later she and her husband bought a motor caravan to explore the Canary Islands, Morocco, Turkey and Europe. Now a widow, Mary continues travel writing for various books and magazines.

Angela Wigglesworth was a general feature writer for national newspapers and magazines before deciding to concentrate on travel writing. She has written three books on small islands – *Falkland People*, *People of Scilly* and *People of Wight*, and now contributes freelance travel and general features to *The Financial Times*. She has three grown-up children and lives in East Sussex.

Index

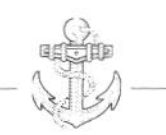

Acknowledgements

The Automobile Association wishes to thank the following photographers and libraries for their assistance in the preparation of this book.

Bryan and Cherry Alexander 160a; **Simon Bannister** 116a, 117, 120, 121; **Barnaby's Picture Library** 138a (D Walker); **Bord Failte – Irish Tourist Board** 16a, 20a, 20b; **Bridgeman Art Library, London** 40 Greek Attic red-figure vase depicting Odysseus with the Sirens, Stamnos, *c*490 BC (British Museum, London); **Bruce Coleman Ltd** 19 (Andrew Davies), 32b (Alantide), 113 (M Freeman), 113b (R Williams), 71 (Dr Eckart Pott), 72 (Mr Johnny Johnson), 151b (Charles and Sandra Hood), 156 (Alain Compost), 157a (John Shaw), 162 (Staffan Widstrand), 164b (G Ziesler); **Comstock Photofile Ltd/1997** 60b; **Crystal Cruises** 82, 83; **Cunard Line Ltd** 176a, 169b, 171a; **Fiona Dunlop** 9a, 150, 151a, 152a, 152b; **Robert Essel** 98b; **Mary Evans Picture Library** 57a; **Helen Gane** 34a, 35, 36a, 36b, 37a, 37b; **Robert Harding Picture Library** 131b; **Christopher Hill Photographic** 16b (Jill Jennings), 25a (Jill Jennings); **The Hutchison Library** 84b (J Henderson); **Images Colour Library** 27a; **Impact Photos** 131a (Keith Cardwell), 132 (Michael Good); **Leisure Cruises SA** 50a; **Magnum Photos** 137a (P J Griffiths); **John Miller** 84/5; **Noble Caledonian Ltd** 22b; **Norwegian Cruise Line** 78b; **Norwegian Tourist Board** 14a; **John Peachey** 15; **Pictures Colour Library** 23, 32a, 43, 46, 54a, 56b, 60a, 63, 64/5, 66/7, 67, 112a, 112b, 172; **Bertrand Rieger** 148a, 148b, 149, 153; **Richard Sale** 98a, 101a, 102a, 102b, 103; **Anthony Sattin** 45, 47a, 47b; **Scott Polar Research Institute** (Bob Headland) 160b, 161, 162/3, 164a, 164/5; **Silversea Cruises Ltd** 155a, 155b; **Spectrum Colour Library** 10a, 10/1, 33a, 42/3, 44, 55, 58/9a, 75b, 86a, 86b, 166b; © **The Stock Market** 62 (Michele Burgess 1995), 65b (Lance Nelson); **Ann F Stonehouse** 99, 100a, 100b, 101b, 101c; **Mary Tisdall** 122a, 123, 124a, 124b, 125a, 125b, 125c, 126, 126/7, 127; **TRIP** 50b (A Tjagny-Rjadno), 51 (V Larionov); **Venice Simplon-Orient Express Ltd** 122b; **The Vintage Magazine Co Archive** 4, 8a; **Angela Wigglesworth** 118a, 118b, 119; **Zefa Pictures Ltd** 12, 13a, 13b, 14b, 28a, 28b, 31, 34b, 54b, 56a, 57b, 58/9b, 61, 65a, 66, 68b, 69, 76/7, 114/5, 115, 157b, 158, 158/9, 159a, 168a, 168b, 169a, 170.

All remaining photographs are held in the Association's own photo library (AA PHOTO LIBRARY) with contributions from:

A Baker 29, 30, 33b; **P Baker** 74, 77, 78a, 79, 171b, 173a, 173b; **P Bennett** 81; **L Blake** 17; **J Blandford** 25b; **C Coe** 59, 106b, 109; **B Davies** 8/9, 88a, 88b, 89, 90a, 90/1, 91, 92, 92/3, 93, 94, 95a, 95b, 96a, 96/7, 140a, 140b, 141, 142, 143a, 143b, 144, 145, 146/7, 147, 154a, 154b; **M Diggin** 18b; **E Ellington** 26a, 26b, 27b; **T Harris** 42; **R Holmes** 87; **P Kenward** 110a, 110b, 111, 159b; **A Kouprianoff** 2/3, 129, 130a, 130b, 134/5, 136, 136/7, 138b; **D Lyons** 75a; **E Meacher** 8b; **I Morejohn** 133, 134a, 138/9; **K Paterson** 48a, 48b, 49, 52a, 52b, 53; **J F Pins** 68a, 70/1, 70, 73; **L Provo** 80b; **C Sawyer** 38a, 38b, 116b, 166a; **R Strange** 104a, 104b, 105, 106/7, 106a, 107, 108; **J A Tims** 39, 41; **P Trenchard** 24a; **W Voysey** 24b; **P Zoeller** 21.